Prai
The Human Race to the Future

"The 21st Century will almost certainly see more technological change than any century before. The computer scientist Daniel Berleant offers a concise, readable guide to what might come next, filled with specific recommendations on what we all need to do to prepare."

— Gary Marcus, PhD, New York University; best-selling author of *Guitar Zero*; author of articles in *Wired*, *Science*, *Nature*, and other major popular periodicals and scholarly journals

"I read through much of the book while teaching for the International Space University in Australia. It is … a thought-provoking enterprise of the first order."

— Joseph N. Pelton, PhD, Director of the Space and Advanced Communications Research Institute (SACRI) at George Washington University; founder and Vice Chairman, Arthur C. Clarke Foundation; Pulitzer Prize nominee

"Berleant takes the reader on an elegantly written, thought provoking ride that could inspire high school students to be more creative, their parents and grandparents to be more future aware, and leaders of nations around the world to take heed and act."

— Donald Maclean, MBChB, retired family physician, psychiatrist, and founding department chair at the University of Illinois

---

A clear alternative to Wesch's "The end of wonder in the age of whatever"… the reader will share in the excitement and joy of this integration of futurism and popular science. According to the author, "My goal is a view of the future at a range of topics and time scales, based on a foundation of science. In melding futurism and popular science, I have worked to illuminate the significance and the beauty of that synergy."

The Human Race to the Future

*What Could Happen —
and What to Do*

The Human Race to the Future

What Could Happen — and What to Do

Daniel Berleant

A Lifeboat Foundation Book

Published by the Lifeboat Foundation
lifeboat.com

—

Spiral clock cover design: Catherine Asaro, with clock art by Francesco
De Comité © 2007 (www.lifl.fr/~decomite)
Circuit board clock, highway, & moon colony covers: J. Daniel Batt
Alien landscape cover: Frank D. Smith

—

3rd edition (November 2015; updated February 2016)
Manufactured in the United States of America
39 41 43 45 47 46 44 42 40

—

Library of Congress Motivated Cataloging Data:
Berleant, Daniel.
The human race to the future : what could happen — and what to do /
Daniel Berleant. — 3rd ed.
p. cm.
1. Science—forecasting. 2. Technology foresight.
3. Science and technology—social aspects. I. Title.
Includes bibliographic references and index
QA175.5.B394 2014

—

ISBN: 0692482768; ISBN-13: 978-0692482766

—

Facebook group: facebook.com/groups/thehumanracetothefuture
Publisher's book site: lifeboat.com/ex/book

—

The Washington
Academy of
Sciences is at
washacadsci.org.

To Joy and Our Next Generation

Table of Contents

Who doesn't have at least some curiosity about the future… about what things may be like some day? This book suggests some answers, spanning from the current century nearly to eternity itself. This book is for you if you are interested in the future, intrigued by science and technology, or both.

The First Generation:
The Next Hundred Years

Labor productivity is so high we could all in principle live comfortably working much less than we currently do. What does this mean? Will we work fewer hours in the future than we do now?

The ancient Greeks used honey water (which they called hydromel) to enhance mental performance. Modern society presents other possibilities, from pharmaceuticals to deep brain stimulation. The future will provide more. *Caveat emptor.*

Keyboards are the classic way to interact with computers. However, general use computers that interact directly with the electrical activity of the brain are getting closer with each passing year.

Chapter Nine (PAGE 64)
Soylent Spring

Artificial meat can in principle be cheaper than animal meat. So, when we finally figure out how to make it taste as good as ordinary meat but at a lower price, society will switch in a big way. There will be no going back.

Chapter Ten   (PAGE 69)
The Turbulence of Short-Term Action

Society is stuck in a short-term mindset, but we shouldn't be. Oddly enough, picking the best actions of an organization or society with respect to a far off end point provides better average yearly performance than optimizing each year individually, one at a time.

Chapter Eleven (PAGE 80)
Battle for the Mind*

Can human memory be edited, revised, and changed? The evidence is in, and it can. Details range from frivolous to advertising applications to the collapse of the destructive "recovered memory" school of therapy. In the future, methods for revising human memory will improve. This should give us pause.

From the title of a book by W. Sargant, 1957

Chapter Twelve   (PAGE 87)
Will Artificial Intelligence Threaten Civilization?

You've heard the story: Intelligent robots try to take over the world. But this is only one possible future. Other futures are explained. One thing seems nearly certain: Artificial intelligence technology will continue to improve, and winning chess and *Jeopardy!* is only the beginning.

Chapter Thirteen   (PAGE 94)
Deconstructing Nuclear Nonproliferation

Nuclear weapons are spreading to more countries over time. Where can we expect this trend to lead?

The law of exponential population increase means that even a small Martian colony could expand rapidly, overpopulating the entire planet with billions of people in just a thousand or so years.

Global warming, if not controlled, will have major effects for hundreds of years to come. Yet over a much longer term, we're actually in an interlude between glacial deep freezes, and glaciers will most likely descend again.

The bible is often noted as the most widely distributed, best-selling book of all time. Therefore it would be no surprise that in a distant future, if robots have formed their own civilization and humankind is a dim and fading memory, that book would survive, perhaps misunderstood and misinterpreted by robot readers with no real understanding of humanity.

The concept of "flower" was a major paradigm shift in plant evolution long, long ago. Future paradigm shifts are also possible. Evolution might make some of them happen over tens or hundreds of millions of years. Yet human genetic engineering could make amazing plants a reality much sooner. The saying "they don't grow on trees" is exactly wrong.

Foreword to

The Human Race to the Future

By Gil Alterovitz and Brandon H. Lee, Harvard University

Thinking about the future can be an overwhelming and daunting endeavor. Yet who is not thinking or perhaps even worried about the future? We consider it constantly, regarding our career path, family, technology, economy, political climate, and more. It is within our curious nature to do so. In our laboratory at Harvard Medical School, the future is something that our students and researchers are looking forward to. Potential discoveries and research in genomics, networks, and standards can have drastic impacts on society. Scientists question the norms and build methods to revise them. As a result, new knowledge is created. Yet in order to know what questions to ask, we need to have a vision of what the outcome may be.

This book looks not only at the coming future of genomic data and personalized medicine but also at various other topics at different time scales. By looking at the history of discoveries and technology, we can only make educated guesses about the future. The book is organized accordingly, envisioning the evolution of progress spanning the future. Furthermore, it is not just about looking forward to the technologies of the future, but at the problems that may arise. How do we protect ourselves from technology? As more and more powerful technologies are discovered and understood, new dangers and protective measures need to be considered.

As scientists at the Biomedical Cybernetics Laboratory, we see the directions in which genomic research can potentially lead. For example, when we are conducting research regarding genomics via predictive models, significant relationships among genes in particular molecular pathways may be determined. Then, further testing in an animal model via a collaborating researcher is done. Researchers are busy finding important new genes associated with biological mechanisms and disease risks. The concept of personal genomics, where access to patients' genetic information affects treatment plans, is relatively new, yet it is the next phase in genomics research as well as technology — its application. The personal genomics market, described in this book, has begun to gain significant traction as its price points begin to drop and more people are able to sequence their DNA. Sooner or later personal genomics and its numerous applications will become pervasive in healthcare and people's lives.

Knowledge is built upon a collective information base. The biomedical informatics field is a clear example of this. Utilizing knowledge of multiple disciplines, we strive to be able to create innovative new methods that permit us to build networks, find patterns within data, and make inferences about the functions of genes. As an analogy, the interconnected aspect of technologies can be likened to a genetic network or map. Genes control other genes and are often part of complex cascades and processes. Technology can also be understood in a similar way.

The mission of the Lifeboat Foundation, the publisher of this volume, is to prevent and minimize existential risks arising from society's pursuit of increasingly powerful technologies. That helps position this book as a moral, ethical, and responsible discussion about the future. The pursuit of new discoveries requires safety nets, preventive technologies, knowledge, and wisdom for the optimal survival of humanity. Great strides must be made to prevent or minimize risks to that objective. This demands responsible thinking about the future and requires a conscious and critical frame of mind.

As the author suggests in the beginning chapters, we are moving into a new economic era. Knowledge is continually being generated that improve technologies and efficiency. In the case of personal genomics, this will enable the ability to produce, provide, and utilize personal genetic information to inform patients, consumers, and clinicians of potential risks of life-threatening diseases. This knowledge is clearly valuable and offers society the opportunity to reach a higher standard of health for all. Eventually, genetic information will not only be read, but may be modified to allow people to improve in many ways. It is important to note that it is the collective knowledge of genetic information that provides us with insight on human genetic diseases, risks, and other biological factors. Our ability to identify patterns within datasets builds upon previous observations and existing understanding.

In the future, Artificial Intelligence (AI) may lead to intelligent entities that exceed our own capabilities in many ways. "Smarter" machines have already won *Jeopardy!* and chess. Gene therapy is eventually expected to replace "bad" genes or even optimize other genes. Within these pages, Berleant provides clear examples of technologies we utilize daily and imagines next steps transcending what we know as today's norms.

The author has contributed significantly to the scientific community through his research in bioinformatics, inference under severe uncertainty, text mining, and other questions. This book demonstrates his continuing passion for such work and many other possibilities illuminated by science and enabled by technology. Despite being written by a scientist, the book includes easy-to-understand prose that even casual readers will appreciate. Although books about the future might be thought to belong within the science fiction section of library shelves, the author grounds himself in scientific facts and knowledge while keeping a sci-fi feel. Taking what we know now, he draws a picture of the future and the potential new technologies that surpass the many we commonly use today such as cellphones, the internet, and computers. We are confident that this book can expand the horizons of scientists, enthusiasts, and the broader audience.

The imminent and distant futures will be based on the knowledge and technologies that exist today. This book gives us a special peek into the near and distant future — with its wonders, problems, and promise.

Gil Alterovitz
Director, Biomedical Cybernetics Laboratory
Center for Biomedical Informatics, Harvard Medical School
Children's Hospital Informatics Program at the Harvard/MIT Division of Health Sciences and Technology

Brandon H. Lee
Research Associate, Biomedical Cybernetics Laboratory
Center for Biomedical Informatics, Harvard Medical School

Introduction

Why Read This Book?

Who doesn't have at least some curiosity about the future...about what things may be like some day? This book suggests some answers, spanning from the current century to nearly eternity.

"Prediction is very difficult, especially about the future," said famous physicist Niels Bohr, quoting his Danish countryman Robert Storm Petersen.[1] Our journey into the future begins with the next hundred years. Call that century-long time frame the "first generation" of future history. After a baker's dozen or so chapters we then move to the "second generation" — the next order of magnitude after a hundred — the next thousand years. The seventh generation then has a ten million year horizon, the very distant future. Beyond the seventh generation are time horizons even greater. This "powers of ten" scaling of future history was first used by renowned physicist Freeman Dyson in his 1997 book *Imagined Worlds*.

Interesting certainties about the future are hard to nail down. Yet the world and our universe will continue to exist (probably!), and new and strange things will still happen. Surely this is a topic worthy of wonder. It is for me. That's why I wrote this book. For future times within your lifetime, predictions can be personally relevant. For horizons farther out, answers are relevant to your descendants, as well as nations and other institutions great and small. For the most distant time horizons, predictions can have cosmic and existential meaning.

Although the chapters are grouped by "generation" or major time horizon, the discussions of many topics actually span multiple time horizons. In those cases the corresponding chapters are placed with the most relevant horizon. This is a matter of judgment that is, at best, inexact.

The future will contain interesting changes in many different areas, so a book like this must cover considerable ground. Thus the topics vary greatly, from computers to biology and numerous others.

This book is for you if you are interested in the future, intrigued by science and technology, or both. Read on, and enjoy the journey.

About the 2ⁿᵈ edition

The biggest challenge in preparing this book for the 2ⁿᵈ edition was getting through peer review by the Washington Academy of Sciences. Every scientist whose work is exposed to serious, anonymous peer review understands pain and frustration; rarely does one agree with the reviewers 100%, and rarely do they leave one's work (and ego) unscathed. Readers will benefit, however, as updates have been made throughout.

...and now the 3ʳᵈ edition

The passage of time continuously converts the unknown future into recorded history, narrowing the cone of uncertainty about what will happen next. "The future is not what it used to be."[2] Two new chapters have been added and previous ones have been updated, but the job is never done. Fortunately in today's world of print-on-demand paperbacks and downloadable e-books, updates are possible within a matter of days, even hours. If you have any update suggestions or notice any errors, please contact us at

- thehumanracetothefuture@gmail.com, or
- facebook.com/groups/thehumanracetothefuture.

The First Generation:

The Next Hundred Years

Chapter One

What It Means That an Hour's Work Yields a Week's Food

Labor productivity is so high we could all in principle live comfortably working much less than we currently do. What does this mean? Will we work fewer hours in the future than we do now?

Everyone eats... but how many of us actually work in agriculture, growing all that food? Way back in 1900, it was 41% of employed people in the US. But by 2000, just 1.9% had agricultural jobs, more than a 20-fold improvement in productivity per person.[1] So if all people wanted was to be fed, the average person would need to work only 1.9% of a 40 hour work week. That's less than one hour a week! *La dolce vita* ("the sweet life")[2].

Realistically, we need additional work time to account for manufacturing the agricultural equipment and supplies as well as distributing and selling the food. However we're still talking about a pretty undemanding average work schedule. Even adding in the working hours required to produce the other necessities of life — water, shelter, clothing, medical care, even jet fighters if you are so inclined — it's still clear that our society is spending a lot of work hours on things we want, rather than actually need. (To be fair, many people do work full time just to make ends meet because the goodies are distributed unevenly.)

The people of the developed world may in the future decide to work less, because productivity is increasing and will eventually become dramatically greater, and free time may become a higher priority than ever more and fancier possessions. This was foreseen as early as 1930, by father of macroeconomics John Maynard Keynes in his essay *Economic possibilities for our grandchildren*.[3] Actual examples have been found in pockets of the labor market, wherein people work less as pay rises. Economists have given it a name: "backward bending labor supply curve."

In fact, in some countries people work significantly more hours per year than others. France famously seems to shut down for a month every summer. The US and Japan don't. In 2013 the Netherlands had an average work week of just 29 hours[4] with 80% of adults employed, vs. 67% for the US. At the same time in the US a typical work week was popularly considered to be 40 hours, although in practice it was often longer. For example, Wal-Mart headquarters typically requires an extra half day of work every Saturday, like Taiwan did in the 1990s. Taiwan reduced its work week to an average 42 hours in 2000,[5] effectively making a typical work schedule include a half day of work on alternate Saturdays (less than the every Saturday for salaried workers at Wal-Mart headquarters in Bentonville, Arkansas).

It is axiomatic that productivity tends to increase over time. Agriculture exemplifies this. It produced, in just 46 minutes per employed person per week in the US in 2000, what in 1900 would have taken 16 hours and 24 minutes. Similarly, the cost of artificial light has also decreased dramatically over time.[6] Manufactured goods, from fabrics to phones, aardvarkiana to zymometers, provide innumerable other examples.

Yet some things are intrinsically in short supply. They do not get inexorably cheaper over time, and thus buck the megatrend of increasing productivity. We simply cannot produce more of them for less. This category includes *land* and *labor* (exemplifying space and time). That means real estate will stay expensive because land cannot be manufactured on a large scale, at least not yet. (On a small scale, yes — Dubai manufactured its Palm Islands, and skyscrapers create new floor space.) Also, the cost of human labor will not become vastly cheaper because work hours are not a manufacturable product. So, economic products that intrinsically consume human work time will also not become vastly cheaper. This includes sports, from Sunday morning football to the quadrennial Olympic Games; live entertainment generally; consulting; and individualized customer, personal, and legal services.

Since time and space will stay about as costly for the foreseeable future as they are now, what does that imply? One prediction is that the traditional economic sector of manufacturing, as it gets increasingly efficient, will become progressively smaller as a fraction of the total labor activity of developed nations. This simply follows the example of agriculture, already extraordinarily small and efficient by historical standards. Counterbalancing that will be services and knowledge work, hallmarks not of the receding industrial age but of the ever-advancing information age and its knowledge economy. This megatrend is illustrated by the transition in developed countries from manufacturing based economies that ruled the industrial age to information age economies based increasingly on movement of information rather than material objects.

The information age is just one illustration of the megatrend from agricultural and manufacturing economies toward service economies, a trajectory that will likely continue, perhaps indefinitely. More generally, economic products that intrinsically consume people's time in their production will become progressively more prominent.

Suggestions

It is a fact that, economically, modern first-world society could in principle be restructured so that everyone works about one day a week. This is due to the small percentage of all labor that goes to producing and maintaining the necessities of life. Then people could devote the other days to doing whatever they wanted. Realistically this will not happen any time soon for most people, though it sure would be nice if it became a generally available option. Even billionaires think people work too much.[7] Since it is possible, and desirable to so many people, it could be a societal goal.

Why not have work gradually become viewed as optional? Considering the increasingly high labor productivity of agriculture and creation of other life essentials, indeed full-time work already is optional in principle. It is not (at least not yet) in practice because of the social and legal traditions and institutions that govern modern society. Yet, while making work optional may seem like a dream from the left side of the political spectrum, it has been a dream from the right side as well. For example Nobel prize-winning, free-market economist Milton Friedman, who advised both Ronald Reagan and Margaret Thatcher,[8] proposed a negative income tax approach to guaranteed income. Richard Nixon attempted (but failed) to implement a plan based on Friedman's.[9]

While societal structures tend to persist, sooner or later they can change. People may want a two-level economy, in which little or no work is required if all one wants are the basics of life.

One of the better known tests of this concept was the Dauphin, Manitoba "mincome" (minimum income) experiment (1974–9). It was found to have several social benefits, while only reducing work time by a little, mostly among new mothers and teenagers.[10] Another step in this direction is Alaska's Permanent Fund Dividend, a yearly (1982–present) payment to its residents from production of Alaska's (i.e., their) oil. The 2015 payment, for example, was $2,072 per man, woman, and child. On a US national level, the Economic Stimulus Act of 2008 directed citizen's dividend-like payments of a few hundred dollars per taxpayer.

In a dramatic but ultimately quixotic step, proponents forced a national referendum in Switzerland on whether to amend its constitution to provide a guaranteed, modest, but livable basic income to its people. On June 5, 2016, the historic referendum failed. But with concern worldwide about automation replacing employment, together with the need to support unemployed people somehow while using them to provide the market necessary to keep economies humming, interest in basic income-like strategies continues.

Currently, not working is an option only for some people. The nonworking poor are supported, rather meagerly, by those who do work and whose work ultimately also pays for disability, unemployment insurance, etc. This is often a contentious topic in public discourse. The nonworking rich are also nicely supported by the very same workers, whose work ultimately enables investments to produce income, businesses to operate, and accumulated wealth to have purchasing power. Of course, there are also retirees, children, and spouses who do not work. Thus the practice of not working is actually quite widespread.

A two-level economy in which life's basics are essentially free and work is essentially optional would dramatically increase our freedom. Many people will choose to work, either for its intrinsic satisfactions, or because they want things beyond the basics that working could make affordable. Many of them will opt for a greater or lesser amount of part-time work. Many will opt for very fulfilling work of their choice: pursuing a dream, say, or raising their families. Others will prefer the luxury of not working. They won't feel like it. (Of course some things need to get done, so if *everyone* goes that route things will fall apart unless robots are developed to do them. Perhaps everyone could be required to work one day a week, or work more but at duties they choose.) These options are economically possible now and in the future could actually be made available to all. The freedom to work as desired — for many, a utopian dream come true.

Chapter Two

Smart Pills'n Such — Cognitive Enhancement the Easy Way

The ancient Greeks used honey water (which they called hydromel) to enhance mental performance. Modern society presents other possibilities, from pharmaceuticals to deep brain stimulation. The future will provide more. Caveat emptor.

People have long used peyote cactus, diviner's sage, and other plants to gain access other planes of awareness (hence the old saw, "Reality is for people who can't handle drugs"). Following the discovery of chlorpromazine back in 1950, pills have also revolutionized the treatment of brain function among the mentally ill.

But what about enhancing reality in healthy people? Drugs that enhance normal mental functioning are becoming better known and, increasingly, more used. We are in the midst of a burst of innovation that, as the future unfolds, will progressively enhance normal mental abilities.

A chemical called alpha-CaM kinase II causes the act of remembering a bad memory to erase it — at least in mice.[1] Sort of like clicking "delete" on a computer to remove a file. That's interesting, though deleting is not the same as enhancing. However there are chemicals that improve memory rather than cause forgetting (Table 1). Remembering is just part of learning, which other chemicals also enhance, at least under test conditions. The common anti-diabetic drug metformin, for example, can make mice learn water mazes better; they even grow more neurons in the process.[2] While any effects on human intelligence are still speculative, it does stimulate human neuron growth in a lab dish. Pharmacological approaches to cognitive enhancement include those listed in Table 1.

Substance, etc.	Main action	Mediating mechanism
Nicotine	Increase concentration (etc.)	Increase acetylcholine (etc.)
Strychnine	Stimulant	Block glycine receptors
Caffeine	Stimulant	Decrease tiredness
Hydromel (unfermented)*	Increase available energy	Raise blood sugar
Sage	Improve memory	Inhibit cholinesterases
Chewing gum	Improve memory	Uncertain
Modafinil (Alertec, Provigil)	Maintain alertness	Anti-sleepiness
Alcohol	Enhance creativity	Improve incubation phase
Methylphenidate (Ritalin)	Improve focus	Stimulate central nervous system
Amphetamines (Adderall)	Improve focus	Stimulate central nervous system
Piracetam	Enhance cognition	Increase brain metabolism
Hydergine (ergoloid mesylates)	Enhance cognition	Increase brain metabolism
Donepizil (Aricept)	Improve memory	Inhibit cholinesterases
Cortactin	Enable neural plasticity	Circumvent calcain biomolecules
Magnesium threonate	Enhance learning & memory	Increase synapse plasticity
Insulin-like growth factor 2	Improve memory	Enhance memory consolidation
Metformin	Improve learning	Stimulate neurogenesis
Exercise	Enhance cognition	Stimulate neurogenesis
Afternoon napping	Improve learning	Clear hippocampal input queue
NgR1 antagonist	Enhance neural plasticity	Reduce demyelination; increase synapse turnover
Fish, except fried	Uncertain	Increase brain tissue volume
Klotho	Enhance cognition	Side effect of life extension
GLYX-13	Improve memory	Acts on brain's hippocampus

Table 1. Some cognitive enhancement substances and activities. (Not intended as a sole source of information. Further readings about rows of table are listed in the References section.)[3]
*Honey (-mel) mixed with water (hydro-).

There are basically only a few strategies for cognitive enhancement to draw from:

- Stimulation. Coffee or any other stimulant makes the brain, or parts of the brain, work harder.
- Neuroplasticity. Learning requires updating the neural connections, so ways of making these connections change faster can speed learning.
- Augmentation. Brain augmentations could work via anything from chips inserted in the brain to hats with embedded electrodes.
- Bootstrapping. If you learn something, then you are smarter. Called "education," this is not the "easy way" to enhanced cognition — but in the long run may be the most effective.
- Health care. Innumerable health related issues, mental and physical, can reduce cognitive performance. Fixing them will thus improve cognition.

Let's look at some examples next.

Transcranial Electrical Stimulation (TES)

Why bother with pills'n such if another method could be more targeted, more specific, and faster? Electricity applied directly to the head! This is not electroshock therapy, which applies hundreds of times more electric current (usually 800 milliamps, more than a typical lightbulb, compared to the 1–2 milliamps typical of TES);[4] TES comes in different varieties: tDCS, tACS, and tRNS (transcranial Direct Current, Alternating Current, and Random Noise Stimulation). Things have come a long way since 1883,[5] when Sylvanus Thompson described connecting a battery to the forehead to cause a "wild rush of colour" (perhaps wisely he did not claim to have tried it himself).[6] For example tDCS was shown in 2011 to be able to induce insight in tricky problem-solving situations.[7] Numerous other studies have shown TES to be able to improve performance in normal people on a wide variety of cognitive tasks related to language, arithmetic, learning, skill acquisition, and planning. You can get TES devices commercially with a doctor's prescription. You can even build your own TES machine on the cheap from do-it-yourself plans.[8] To control where the electricity goes in the brain, a device can simply permit appropriate placement of the electrodes on a headband or cap (a "smart hat").

However it may be better to administer electricity to the brain in a more focused way.

DBS

Taking the "focus" idea to a seeming extreme, in February 2007, a man named Sean Miller, suffering from stubborn depressive disease, had electrodes surgically implanted directly into his brain. The treatment was successful. By 2009, tens of thousands of people had been treated for another disease, Parkinson's, using brain-implanted electrodes. It has also been shown useful in treating obsessive-compulsive disorder (OCD). The procedure is called Deep Brain Stimulation (DBS), and there is no reason this technology has to stop at treating disease. Using it instead for brain enhancement seems like, well, a no-brainer. Indeed, in September 2011 DBS finally made mice smarter.[9] The electrical stimulation caused the mice's brains to grow more neurons, and long after the electrodes were removed the mice performed better in water mazes. Humans, we runners of the maze of life, not to mention of the rat race, may willingly be next.

Like selected psychoactive drugs, DBS could potentially enhance cognitive performance in healthy individuals. It would be important to find the exact and correct location in the brain to stimulate to, say, motivate yourself for an hour, do better in mazes, such as corn (i.e. maize) mazes, feel more romantic, or remember where you put the darn cellphone, of course. On the other hand, who knows what would happen if you accidentally stimulated the wrong place? You might end up demotivating yourself for a week, standing up a date, or remembering where you last saw the cat (which has long since moved to a sunnier spot).

Luckily, progress in brain scan technology is improving exponentially.[10] We'll soon know a lot more about where in the brain it does what.

TMS

Some people will always be curmudgeons who balk at jabbing wires through their skulls, no matter how much they are reassured that it is "harmless." That's a reason third generation brain stimulation technology is so intriguing (first generation — pills; second — electrodes inserted into the brain; third — noninvasive neurostimulation). In addition to TES, described above, one of these methods is *transcranial magnetic stimulation*, or TMS. First approved for fighting depression, the qualitative physics works like this: Electromagnets are positioned on the head so to direct magnetic lines of force through the area of the brain to stimulate. Then the magnets are activated with alternating current, or AC (ordinary household AC alternates, i.e. changes direction, at 60 forward-and-back cycles per second in many countries, for example). Changing direction changes the magnetic lines of force correspondingly. This in turn changes electrical fields in the affected brain tissue, changing the electrical activity of neurons in that area. That, of course, means the brain is working differently.

Ultrasound

Another method uses ultrasound, or sound waves above the audible range of frequencies.[11] Ultrasound is already a standard technology for medical imaging, for example, of a developing fetus in its mother's womb. A different frequency and power is used for brain stimulation. The nanomechanical effects of the ultrasonic sound waves can open voltage-gated sodium ion channels in neural membranes, leading to neural firing.[12] Maybe if you were smarter you could comprehend that last sentence a little faster. To get smarter, perhaps you could use ultrasound, TMS, TES, or some other method to stimulate your left lateral prefrontal cortex. It is near the front of the head, just slightly off to the left side, on the surface of the brain. Its connectivity to other parts of the brain explains 10% of the variation in intelligence in humans.[13] While you can't change the connectivity you have (yet), stimulating what you do have there might just make you smarter.

Minimizing invasiveness

No one wants to drill holes in the skull and jab electrodes into the brain to do something that could be done less invasively with TES, TMS, ultrasound, or some other even less invasive method. For example, some brainwave frequencies are more associated with effective cognition than others. By displaying to the user their brainwave frequencies in real time they can actually learn to change them as desired,[14] a process often called neurofeedback. Ordinary flashing lights and rhythmic sounds can also be used to modify brainwave frequencies. This is called brainwave entrainment (BWE), audio-visual stimulation (AVS), and audio-visual entrainment (AVE). Some evidence suggests that this can "enhance attention, increase overall intelligence, relieve short-term stress, and improve behavior."[15] Good news if true.

Visual hemispheric invocation (VHI) has been proposed as an equipment-free way to instantly access the powers of the right or left hemisphere. To invoke the right hemisphere, simply focus your eyes on a point, while focusing your attention on something to the right of that point in the peripheral visual field. Then observe any changes to your emotional or cognitive state. What could be simpler than that? This process is based on the fact that the right visual field excites parts of the left hemisphere (and vice versa for the left visual field).

Smarter kids

People want their kids to be smarter. Kids would benefit. Even small improvements to IQ correlate with significant increases in lifetime earnings — roughly $15,000 per IQ point in the US (in year 2000 dollars).[16] Worldwide, "Iodine deficiency ... can significantly lower the IQ of whole populations"[17] and can be alleviated by iodizing table salt in affected countries at extremely low cost, according to the Micronutrient Initiative, a Canadian non-governmental organization (NGO).

Education is the classical approach to increasing intelligence. Of course, that's hard work, so shortcuts are always of interest. Playing classical music to a fetus still in the womb may be fun for mom, but as an intelligence enhancer for the upcoming little one, its usefulness is highly speculative. However, taking choline dietary supplements during pregnancy makes for smarter offspring in rats.[18] Quite possibly it would work for people, too. Choline is a nutrient found in egg yolks, lecithin, and other sources. When women who are pregnant or lactating take cod liver oil supplements, their children are smarter at 4 years of age, which is good, but any such effect is no longer clear by age 7 — not so good.[19]

Longer term, controlling the genes of our children to increase their intelligence will become more feasible. Technologies include genetically modifying sperm and egg cells, generating multiple embryos in the lab and implanting only the smartest one in the womb, and abortion based on analysis of embryonic DNA. Some people will not do these things, of course, but others will; some countries and religions will not allow it, but others will. Success will depend on identifying which genes help control intelligence. As of this writing, no single gene is known to have a great effect, but some appear to have small effects and others have effects yet to be discovered. Many genes with small effects may add up to a large effect. Thus, it will soon become feasible for biostatisticians to determine the influence on intelligence of every normal variation of each of our 20,000-odd genes. At that point medically engineered superbabies, built for their mental abilities, will be in sight.

What to do

First, be careful. The risks of abuse, unanticipated side effects, and long-term, delayed effects should be a concern. Even honey water (hydromel), over a long period, might increase chance of diabetes through repeated blood sugar spiking. Yet the competitive disadvantage of non-use could also be a problem in some situations. Supporting anti-doping efforts could help solve that problem: Inasmuch as we are discussing becoming smarter, the term anti-"doping" is apropos. Avoid harming yourself; no need to be a dope.

A tried-and-true method for cognitive enhancement that, should be sought, is to associate with others who have similar goals and interests. Sometimes called the "critical mass" effect, anyone who has been in such a group can vouch for its effectiveness.[20]

Chapter Three

Keyboards Yesterday,
Mind Reading Tomorrow

Keyboards are the classic way to interact with computers. However, general use computers that interact directly with the electrical activity of the brain are getting closer with each passing year.

Computer users once handed stacks of cards to a technician for overnight processing. Computer keyboards and mice are better. Yet they too seem increasingly awkward and outdated, especially as very small computers in the form of smartphones are becoming so common. Future users will prefer much more convenient methods.

Driven by the need for ever-more efficient computer input, keyboard use is not-so-gradually becoming replaced by speech understanding. Advances in speech understanding are increasingly found in everything from head-mounted computers to telephone answering systems. But direct brain-to-computer communication — mind reading — will soon follow into the mainstream. Speech input will reach its potential first because it is a fast-maturing technology. Mind reading is still in relative infancy, though advancing quickly.

Progress during the previous century severely lagged the hype, leading to widespread disappointment and lack of research funding. Things began to pick up in the current, 21st century. Early systems asked you to say "yes" or "no," for example, or a digit from 0–9. Yet they are distinguishing ever-larger and more complex verbal commands, to the point that by 2015 smartphone assistants like Apple's Siri were understanding arbitrary short verbal commands well enough to be useful. So excitement is rising and advances continue.

There is now ample reason for optimism and little for pessimism. The supporting hardware, linguistic knowledge, database, and software algorithm technologies march on. There seems to be no intrinsic reason why computers will not advance beyond even human's own speech performance. Thus the human capability for speech understanding will first become within reach, and then exceeded, such as by the adding in the ability to understand lots of foreign languages.

The impetus is that speech is more convenient than typing. After all, people learn to talk without apparent effort at an early age. While typing may seem effortless with sufficient practice (and a good keyboard), it actually is a tediously learned skill that benefits from frequent nail-trimming. A microphone for speaking is both smaller and cheaper than a keyboard for typing. Besides, smartphones already have microphones. Furthermore, speech is easily proven to be more powerful than keyboards: Any keypress can be verbalized by saying the name of the key, but many verbalizations have meaningful nuances, tones, and other sounds that keypresses cannot capture. Thus even though keyboard technologies will get progressively cleverer (e.g. the projection keyboard, which consists of nothing but a lighted image of a keyboard projected onto the nearest surface, as of 2015 made by Celluon, Inc.), they will probably still fade away.

But why stop with speech?

Mind reading

Mind reading technology in basic form is already here. It will continue to improve because of the need for ever more convenient computer input.

Amazing things can be done by sticking electrodes directly into the brain (video: tinyurl.com/ncc3nxw). However, let's focus next on examples that work from outside the head and do *not* involve inserting wires inside, on the grounds that most people will stubbornly prefer it that way.

A brain wave reading device that enables paralyzed individuals to switch lights and appliances on and off without moving at all was reported ready for marketing as early as 1997.[1] This device, the MCTOS Brain Switch, was still listed for sale as of 2016.[2] In 2008 a team of researchers at Keio University,[3] Japan, enabled a paralyzed man to stroll about on Second Life, a computer-simulated world, by outfitting him with a brain wave detector and software that read his thoughts about limb motion. By 2013, non-invasive brain wave detection enabled people to control quadrotor air vehicles by imagining hand motions.[4] Beer finally entered the picture in 2015 when Erik Sorto, paralyzed from the neck down, drank some using a robot arm controlled by his thoughts.[5] How long will it be before the first driver's license is issued to a quadriplegic, or before brain wave reading smartphones no longer come with — what did they used to call them? — "keypads."

Keypads and voice recognition systems are actually mind reading devices, just ones that require translating brain impulses into finger or vocal chord movements as an intermediate stage. It is more efficient in principle to read brain impulses directly, dispensing with the intermediate stage of transducing nerve impulses into finger and vocal chord movements. For the number keys, the concept was first demonstrated in 2009,[6] and will only get better over time.

As with computer speech understanding, demand should drive incremental progress in the number and variety of thoughts that are readable by mind reading computers. In 2008, fMRI (functional magnetic resonance imaging) was used to observe brain activation patterns associated with thinking about nouns. A computer was trained to recognize these patterns with "highly significant accuracies"[7] for 60 different nouns. A similar capability was also demonstrated with viewed pictures rather than imagined words,[8] because these similarly activate concepts deeper in the brain. From that somewhat humble start, improvements continue.

Commercialization is a prime incentive of technological advances, and computer gaming provides a ready market for mind reading hardware and software. In 2008 NeuroSky Inc. launched the first mass marketed, brain wave-sensitive headset. Regarding the price-performance of mind reading technology, according to Emotiv Systems, another company, "There is no natural barrier from what we can see."[9] In 2015 another company, OpenBCI (for **Open** source **B**rain **C**omputer **I**nteraction) began not only marketing brain wave reading system components but also promoting a do-it-yourself maker community to crowd source innovative applications and improvements.

Military applications have historically also driven technological advances. For example, ancient catapults used springs made from elaborately constructed cables of...human hair![10] A little more recently, the US Army sank millions of dollars into developing technology for a future "thought helmet." It was to read soldiers' thought messages and transmit them by radio to fellow soldiers whose receiver equipment would convert them into a speech feed. Soldiers would need to think in "clear, formulaic ways ... similar to how they are already trained to talk."[11] Civilian applications for this sort of technology abound (although the implications of such devices for communicating with, say, a romantic partner in "clear, formulaic ways" like soldiers on a battlefield remain to be fully explored). For example, in 2010 an off-the-shelf brain wave reader was demonstrated as part of a cellphone device that could detect who the user wants to call, and dial the number, though physical movement (a wink) as a trigger helped.[12]

Other interesting advances in mind reading are also occurring. For example in 2009 a technique for reading the mind and outputting a fuzzy movie corresponding to a movie the subject was actually viewing was described.[13] On a different note, a 2013 project read interpretations of Beethoven from peoples' brains and played the results using a full orchestra at the Peninsula Arts Contemporary Music Festival.[14] As mind reading improves, lie detection will never be far behind. The No Lie MRI company was providing limited commercial services in 2014 with ambitious expansion plans. They analyze brain scans of people under questioning.

Progress seems set to continue for the indefinite future. The necessary mind input hardware devices should get progressively lighter, more portable, and more comfortable until they are as easy to wear as baseball caps or sunglasses — and as cool (or even cooler). In fact, they may well *be* baseball caps or sunglasses, with the necessary electrodes built right into the hat band or earpiece. For those who don't want on-off switches on their ball caps, but do want to hide their thoughts from their mind reader app at times (since who wants every thought read continuously?), the caps could be built to provide another on-off method. Just wear it backwards to turn it off, frontwards to turn it back on. For women who don't care to wear their baseball caps backwards, no problem. They could be made to turn off by putting your hair through the back and on again keeping the hair under the hat band.

Recommendations

Many of us would like to know how fast mind reading technology is advancing in order to know when to expect specific capabilities to arrive. To assess the rate of progress and estimate future advances, technology foresight specialists could be commissioned to better quantify the state of mind reading technology and its rate of progress.

Various "laws" describing rates of advancement of various technologies have been proposed. The most famous is Moore's Law, describing the number of transistors on microelectronics chips.

Various others have been proposed over the years as well, such as Kryder's Law (about increasing memory capacities), Hendy's Law (on improving digital camera resolutions), Butter's Law (describing rising data flows in optical fibers), Nielsen's Law (concerning escalating home network speeds — let's just say I have concerns about that one), and Carlson Curves (describing advances in biotechnology). In addition Wright's, Goddard's, and hybrid laws apply generally to many technologies.[15] An exponential trajectory will generally slow down given enough time, revealing itself as merely the initial part of a "logistic" or S-curve. *[News – Feb. 9, 2016. Moore's Law, evidence suggests (goo.gl/xoYVra), after many decades is now entering this phase.]* This slowing in turn can be the first stage of an even longer-term curve that ascends, levels off, then descends (think landline telephones, mechanical clocks, and horses). Thus I propose a law, too: The fraction of all the exponential laws ever identified that are still in force declines over time.

Getting back to brains, spatial resolution of non-destructive brain scan technology doubles about every six years, while time resolution doubles about every year and a half.[16] This forms an exponentially improving foundation for future brain reading technologies. A good metric for progress in brain reading would also include the volume of brain tissue scannable at a given resolution. Adding even more factors could likely result in a smoother and more compelling exponential trajectory than science, engineering, or sales figure components alone. After all, progress does combine scientific, engineering, and sales factors. Even the seemingly simple Moore's Law (number of transistors on a chip) actually combines multiple factors (e.g. transistor size, chip kernel (die) size, number of layers, and layout algorithms). While awaiting the discovery of a great metric for progress in brain reading, I guesstimate that high-capability speech interfaces will start giving way to mind reading devices as a cheap, widespread commodity fairly soon.

Chapter Four

Wiki-Wiki-Wikipedia

Wiki-wikipedia, an extension to Wikipedia, could provide a view of any article from the "colored lens" of any other article automatically. People could edit those automatically hybridized articles for use later by others, just as they do with Wikipedia's current articles. Yet the short passage concept demonstrated a hundred years ago by Paul Otlet's Mundaneum could extend Wikipedia even further, from Wiki-wikipedia to Wiki-wiki-wikipedia.

The web (birth name: World-Wide Web) is but the most recent communications revolution. Others stretch back in time: television, radio, telephone, the telegraph, newspapers, the printing press, writing, and human speech. These innovations transformed society. They also made us better informed and thus (we hope), smarter.

Genesis of the web

More than earlier disruptive communications technologies, the web brings to fruition a long-standing dream of information connectedness. According to web inventor Tim Berners-Lee et al. in 1994,[1] "The World-Wide Web (W3) was developed to be a pool of human knowledge."[2] The web swept forward and soon "overflowed this original goal to become a vast sea"[3] as two commentators put it. Sir Tim was knighted by Queen Elizabeth II in 2004. But the dream long predates Berners-Lee and his lab at CERN, the storied Swiss birthplace of not only the web but also the Large Hadron Collider, the world's most powerful atom smasher, or particle accelerator.

Earlier visionaries also foresaw the potential of general, customized access to the vastness of the world's information. One was Douglas Engelbart (1925–2013). Engelbart, in the 1960s, built the first hypertext system,[4] in which text contains links to other texts — the hyperlink idea that makes the web possible (Figure 1). What about clicking those links? Engelbart also invented the mouse![5] These advances grew from his first day back to work after getting engaged in 1950, when he had an unnerving epiphany, visualizing his future career as a "long, long hallway, ... almost featureless."[6] Over a period of months he pondered this until another epiphany occurred: "What if I could contribute something significant to how humanity could cope better with complex sorts of problems? ... BANGO!" But how? Vannevar Bush's famous 1945 *Atlantic Monthly* article, *As we may think*, was the key to solving that puzzle.[7] He got to work and never stopped.

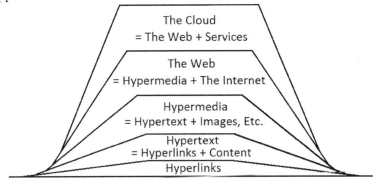

Figure 1. From hyperlinks to the web. Each technology was built on a foundation of its predecessor. What is or will be built on the cloud?

Working independently, in 1965 Ted Nelson coined the term hypertext (but with a hyphen: "hyper-text") at Vassar College in New York.[8] A colorful character, Nelson is reputed to have also coined the term "teledildonics." His Xanadu project[9] embodied his vision of the universe of documents, or "docuverse," in which "The World Wide web is what we were trying to prevent."[10] Yet his ideas are present in the web, for example in the concept of "transclusion," which describes "hot-linking," the standard way of putting images into web pages. As with Engelbart, Nelson was deeply influenced by Vannevar Bush's article *As we may think*, which he reprinted in his own book, *Literary Machines*.[11]

Often credited as the inspiration for the developments leading to the web (mostly by those unaware of Paul Otlet's work, as we shall see), is a device Vannevar Bush conceived in his *As we may think* article (Bush also wrote another article, *It is earlier than we think*[12]). He called this hypothetical device the memex. 'Memex' is often described as either from MEMory and indEX, or MEMory and EXtender. In the article itself, Bush states, "It needs a name, and, to coin one at random, 'memex' will do." But who would really believe that?

A memex would serve as a repository for all of the written materials that one might accumulate over a lifetime, including books, notes, and other things, stored on microfilm to save space as there was no flash memory, DVDs, etc. in those days. Levers and screens would enable one to easily find, skim, and read anything stored in the memex. Bush wrote, "Wholly new forms of encyclopedias will appear, ready made with a mesh of associative trails running through them." Sounds like Wikipedia! A memex will probably never be built and, born in 1890, Bush died in 1974, too soon to even use Wikipedia (or even Webopedia, once pioneering but now just a stodgy, out-of-date also-ran).

The milieu that inspired Bush's "associative trails" of hypertext was the birth of the general-purpose electronic computer. The first was the Manchester Small-Scale Experimental Machine (SSEM), or "Baby" for short. Baby went live in June 1948, but excitement had already been building. The famed ENIAC appeared in 1945, but though a general-purpose computer, was not fully electronic (it was programmed by flipping switches and connecting wires). The Z3 was built in 1941, but it was not electronic at all, relying on electrically driven mechanical parts. The original general purpose computer was the Analytical Engine of Charles Babbage in the century before. It was fully mechanical.[13] Its construction was sadly never completed owing to lack of funds. Programs were nevertheless written for it in 1842–3 by Ada Lovelace (more precisely, Lady Augusta Ada King, Countess of Lovelace),[14] who thus became the first computer programmer.

But before Berners-Lee, before Engelbart, before Nelson, before Bush, and before the computer revolution that took off in the '40s, there was Paul Otlet. Born in 1868 in Brussels, Belgium, to a wealthy family, his mother died when he was three. Despite this hardship, he ultimately developed many of the ideas that make the web a unique, indispensable part of the modern world — but using paper cards instead of computers.[15] His favored unit of knowledge was a passage short enough to be written on a single 3×5 index card. He dreamed of putting the world's knowledge onto these cards and using the resultant card collection as a tool for the betterment of humankind. Writing in 1892, he envisioned a "systematic and very detailed synoptic outline of knowledge" which "would have enormous advantages" such as, for example, "the creation of a kind of artificial brain by means of cards."[16] Starting in 1895, this was implemented as the Universal Bibliographic Repertory, which soon answered questions by mail by sending copies of various index cards for a fee. A collection of over 15 million such cards was ultimately housed in what was named the "Mundaneum," in Brussels. Closed in 1934, its remnants still exist (see www.mundaneum.org).

Recommendations

From Wikipedia to Wiki-wikipedia

The web is a quality-*un*controlled, giant disorganized mass (or maybe mess) of information. But Wikipedia is just the opposite: organized, quality-checked, and much smaller (though still pretty big). So it is no surprise that Wikipedia is one of the best-known and important sites on the web. Together, they complement one another and vindicate the dreams of the visionaries from Paul Otlet on.

Yet Wikipedia can make a quantum leap: It can become Wiki-wikipedia. Wiki-wikipedia builds on the basic fact that Wikipedia users often seek information in *context*. For example, most readers of the Wikipedia article on computers are not going to read the article from start to finish. Their interest has a specific slant — a context — that varies from one reader to the next and even, for the same reader, from one day or moment to the next. If that slant is history, they can find some information on computer history in the computer article by spending (i.e. wasting) time looking for it prior to actually reading it. Worse, if their slant is the future of computers, as of this writing they can spend/waste time looking, yet find nothing.

What is needed is to increase the current set of articles on individual subjects by adding articles on the intersection of any *two* subjects (hence the "Wiki-wiki-" in "Wiki-wikipedia"). If there are 5 million single-subject articles, there would then be 5 million × 5 million double-subject articles. That's 25 trillion! These include computers+history, computers+future, and 24,999,999,999,998 others. Most of these articles would be generated on-the-fly by automatic extraction whenever accessed by a user (e.g. containing passages about history from the computer article, plus passages about computers from the history article). But any of these could be hand-edited by wikipedians just like ordinary single-subject articles, after which they would be stored indefinitely for future access just like ordinary single-subject articles. More importantly, Wiki-wikipedia will be a better home for the many of us who would benefit from not just general information, but information customized to our needs at the moment of access.[17]

Wiki-wiki-wikipedia

Providing Wikipedia users with information in context — Wiki-wikipedia — is a start. But tailoring information to the specific need of a user at a specific moment requires true customization. Taken to its natural conclusion, this requires a system that dialogues with its user, answering a question, receiving a follow-up question, answering that one, and so on, thus efficiently transferring knowledge from system to user. What unit of information should such a system use as the coin of its realm?

The chiseled inscriptions on animal bones that helped prove the existence of the Shang dynasty of over 3,000 years ago were basically sentences. Otlet's 3×5 index cards were too. Tutors interact with their pupils in a highly interactive manner, frequently using sentence-length passages. The Socratic method also relies on sentence-length passages. That is because the sentence is a natural unit of knowledge.

If the future of Wikipedia is Wiki-wikipedia, then the future of Wiki-wikipedia could be a user interface that extends the current Wikipedia interface by permitting a user to engage in a highly interactive dialogue. In this efficient dialogue session, the user asks questions and each answer is intelligently chosen to be the best possible sentence-length passage from the Wikipedia or Wiki-wikipedia body of knowledge. This 3rd generation Wikipedia could be called: Wiki-wiki-wikipedia. This is basically (Wiki-)wikipedia fed into IBM's Watson software and overlaid with Siri for voice input and output, so the first versions should be arriving soon.

Chapter Five

Live Anywhere, Work Anywhere Else

Current commuting and living practices continue in part because we're used to doing it that way. These habits are ripe for change. The economic incentives favor telecommuting and robotic telepresence. Here is what the insistent push of these incentives means for the future of living, working, and commuting.

Think about the poor commuters (maybe you know one of those unfortunate souls?). They must travel to work each day by car or some other way. Cars are expensive. Roads are expensive. Road maintenance is expensive. Fuel is expensive. Time in traffic or, worse, traffic jams, is wasted. Commuting by car has been found to be a debilitating, continuous drain on our happiness and quality of life.[1] Think about it this way: If you save 45 minutes/day in commuting time for 250 days/year, that's 187.5 hours saved every year. That is more than the work time replaced by *a month-long vacation.* If you could spend that commuting time working instead, wouldn't you enjoy that month (or more) of extra vacation time, every single year?

On the other hand knowledge workers, unlike factory workers for example, increasingly can work off-site such as at home offices, and only travel to work occasionally. Bringing your knowledge work home used to mean a briefcase crammed with information printed on rectangular mats called paper. Then floppy disks gave way to small plug-in flash drives. Now storage in the cloud has rendered even the smallest personal physical memory objects unnecessary. Collaboration over great distances is now often possible without travel. An example is computer software development teams that work in shifts. The Americans work a day shift, go home, and while they sleep at night, in India they work their own day shift.

Meetings have progressed increasingly from in-person to all-electronic. The dramatic savings in time and money compared to a plane trip, hotel bills, etc., explain why. The 20th century was littered with videophone projects that were commercial failures, like the AT&T Picturephone that captured imaginations (but not profits) in my youth. However the 21st century is seeing a dramatic trend of increased use of electronic meetings with video. One application of this is education. Distance education in real time is revolutionizing higher education as students can increasingly log in to their scheduled class sessions from home, taught by instructors who also can potentially be at home. When home is far away, maybe another country, electronic education can enable learning that otherwise could not occur at all. When home is merely across town, there can be remarkable savings in debilitating commuting time and gasoline, especially when added up over the people in the class and the many class sessions. As the trend continues, physical classrooms, parking lots, and so on will become less and less extensive. Why build a new building full of large classrooms when it is cheaper just to increase the percentage of students logging in from home? The education sector of the economy is thus becoming more efficient over time, just like other areas of the knowledge economy. Of course, this has negatives like reducing face-to-face interaction and so tends to increase isolation, tending to counter the positives.

As the telecommuting trend continues, where you live will become increasingly unimportant. Live anywhere, work anywhere else! Some feel that urbanization is the future.[2] However, ruralization may win out in the end. Cities provide efficiencies of scale — for example in per capita energy consumption, time, and bringing together creative "critical masses" of individuals — yet the dark horse and likely ultimate winner is instead the web. The World Wide Web is making us all part of a giant global digital village,[3] enabling efficiencies beyond what merely living nearby would or even could provide.

Telecommuting can replace commuting but arguably only to a point. That point can be exceeded by combining telepresence and telecommuting: Videoconference in to work, but with the video hookup going to a telepresence robot. This is a computer on a wheeled stand that you remotely steer here and there, up and down the halls, hither and thither to conference rooms and offices, nearly anywhere you might walk if you were actually there. (Except the cafeteria, since robots don't get hungry.) Such telepresence is now supported by a number of robotics companies. They take off-the-shelf computers running ordinary videoconferencing software, mount them on stands with motorized wheels, and provide control from the remote user's computer.

An important question to many people will be, of course, "What kinds of jobs will become possible to do remotely and which won't?" Jobs that deal with information or communication are relatively easy to make remote. Anything involving sitting in front of a computer is a candidate, from highly paid, skilled system designers to the content moderators sometimes known as data janitors.[4] Anything that involves telephones can also be done remotely. On the other hand jobs that require handling objects (like manufacturing) are harder to make remote. Similarly for jobs that require dealing with people in person. Massage is a task that cannot be done remotely. Emergency medical technicians, police officers, and taxi drivers can't work at home. But their dispatchers could.

Employment without employees

As people increasingly work at home, their homes increasingly far afield from their workplaces and colleagues, we will become a live-anywhere, work-anywhere-else society. The current distinction between full-time and part-time employment will get fuzzy. Workers will increasingly become like small companies selling their services. It will be natural for them to band together in influential professional associations. This will help protect their interests in an employment environment that is increasingly disinterested in them. The "employee of organization X" aspect of our identities will become correspondingly more divorced from our careers, lives, and identities. Labor protections will recede, but voters may decide to demand government social safety net support to offset the dwindling of employer-provided health care, pensions, etc. Some people will have multiple employers at once. Work that improves our employability will be in greater demand; work that does not enhance employability will be less desired and thus need to offer higher pay to get people to do it. Profit sharing schemes will proliferate. Employee-owned businesses will command greater employee loyalty and hence have an edge over other companies. Compensation will become highly renegotiable. Pay levels will both plummet and soar in response to changes in both occupation-specific supply and demand and general unemployment rates.

Recommendation

Working at home is a dream come true for some. Luckily for them, telecommuting will increase. Yet many telecommuters will grow to really appreciate sometimes getting away from home's comforts and distractions. That way they will get more done. Still, few people like commuting through a tedium of traffic jams and lights just to get to work and home again.

If the problem with going to work all the time is that it is preferable to work at home, the problem with working at home all the time is that many people will want to get out a bit more when they work. Hence an increasingly needed business opportunity: office centers that provide their clients — telecommuting employees of other companies as well as the self-employed — with office space that is conveniently close, but not at, home. Small cubicles will rent for a low price, and larger cubicles and walled-off offices for more. A center could offer just a few or potentially hundreds of units depending on the local population. Each could be outfitted with internet connections, technical support specialists on-site (or working remotely from home), shared and unshared printers, coffee machines, janitorial services (provided by people who cannot work at home), food vendors, etc. — anything desired by people with a wide variety of needs and budgets who wish to rent a place to get away from home to work in. Some telecommuters have been using internet-friendly establishments like Starbucks for this for some time already.[5]

Such a center could also provide a sense of community and shared experience to its occupants. Instead of going out to lunch to celebrate Cathy or Bert's birthday in marketing, people would be celebrating Emily or Joe down the hall's birthday *even though the birthday guy or gal works for a completely different company*. Valuable opportunities for networking would also present themselves, because one's "co-workers" will be working for various different employers. This will enable many people to know a lot more than they do now about the diversity of company cultures, salaries and other employment practices, opportunities in their areas, and so on. Clearly, many people would benefit from such a stimulating and informative work environment.

Chapter Six

From Highly Centralized to Highly Decentralized Society

Modern civilization depends on centralized food and energy production as well as the resultant required distribution networks. This is risky, even dangerous. In the future, gardening and farming robots could mitigate the risks of the food distribution system, and solar energy could mitigate the risks of centralized electric utilities. Society will be much more resilient to disaster as a result.

Modern techno-civilization has increasingly become a world network with capabilities previously undreamed of. But modern civilization has an Achilles' heel. Like the mythical Achilles and his famous heel, it is potentially vulnerable to catastrophe, possibly even widespread collapse. That vulnerability is caused by over-centralization, which makes modern civilization so complex and interconnected that catastrophic disruptions may be both unpredictable and inevitable.[1]

For example, many people have never seen an electric generating facility, because these generators are few and far between, and they send electricity over long distances. Yet people depend on electrically operated lights, phones and other electronics, appliances from refrigerators to washing machines to alarm clocks, and equipment from gas pumps to air conditioners. Indeed the average person uses electricity almost constantly and is barely able to function during even short outages. If a large electric grid went down, the first five minutes might be fun. The first five years could end civilization as we know it over a continent.

Fuel

The situation for fuel is similar. By and large, liquid fuel comes from giant oil refineries. No large refinery has been built in the US since the last one began operations in Garyville, Louisiana in 1977.[2] In late 2011, however, two big ones in Pennsylvania shut down. Gasoline goes from the existing large centralized refineries to our cars' gas tanks and ultimately out the tail pipe as water vapor and carbon dioxide (plus other pollutants). Yet most of us rarely even see gasoline. To illustrate, many people do not know its color. Do you? Next time you fuel your car, shake a couple of drops off the end of the pump nozzle as you remove it from your tank, just to remind yourself what the stuff actually looks like with your own eyes (but even that is not enough to show the color, which is light amber). If gasoline distribution was halted, or a few refineries destroyed, this would prevent modern society from functioning in a normal way — a major crisis. If conflict with Iran heats up, for example, expect Iran's main refinery to be neutralized, which alone might bring down the government. Similarly, five years without any gasoline would transform and weaken society if not destroy it completely.

Food

Another source of over-centralization is food. An increasingly tiny percentage of the population in industrialized nations produces nearly all the food. Most people rarely even see farms, but depend on them daily (or more accurately, 3× a day). Getting food from where the farms are concentrated, such as wheat from North Dakota, Kansas, and Montana; corn from the Midwest; potatoes from Idaho; fruits, nuts, vegetables, and other specialty crops as well as nearly all other foods, requires transportation. Even living where food is grown doesn't eliminate the need for transportation — you're much better off living next to a grocery store than in the middle of a corn field. If transportation stopped for five hours, no problem. Five years would, to continue a theme, end modern civilization as we know it.

And so, we depend rather dramatically on centralized production of electricity, fuel, and food. What could cripple these systems and what can we do to prevent it? For electricity, a sudden interruption in fossil fuel would cause a sudden ending of most electricity. There might be enough capacity left from generators powered by nuclear, wind, solar, hydroelectric, geothermal, and other energy sources to save civilization as a whole. At least we can hope so. Fortunately a complete interruption in fossil fuel seems unlikely, though prices might zoom enough to create havoc in the general economy. The reason is that fossil fuel is obtainable in different forms (coal, oil, natural gas) and from many sources, and it is hard to see all of them being shut down at once. In recent times major blackouts tend to be caused not by fuel interruptions but by sudden, uncontrolled electric grid breakdowns. The US grid, for example, is an aging distribution infrastructure, increasingly precarious and prone to outages that risk causing regional and temporary chaos. Luckily such blackouts seem unlikely to cover whole nations for indefinite periods of time, so at least we're probably safe from an existential crisis stemming from gradual deterioration of the grid. But gradual deterioration is not the only thing that could go wrong.

Disaster scenarios

An electromagnetic pulse (EMP) occurs when magnetic fields change quickly where wires are present, destroying unprotected equipment, from electric grid control units to TVs and other home electronics, by causing electric surges in the wires. The underlying physics behind these out-of-control surges is the same as the physics that enables the electricity produced on purpose, by huge coal-fired generators and small automobile alternators alike. An EMP can occur as one effect of an air burst of a nuclear weapon, as documented historically in both US and Soviet weapons tests. According to congressional testimony,[3] "The electromagnetic pulse generated by a high altitude nuclear explosion is one of a small number of threats that can hold our society at risk of catastrophic consequences."[4]

The physics of changing magnetic fields around wires also explains the dangers of severe geomagnetic storms. These occur when disturbances in the solar wind (a fast-moving flux in space of mostly electrons and protons from the Sun) in turn cause erratic movements in the Earth's magnetic field. Under normal circumstances this field is easily detected with an ordinary compass, is all around us and, more to the point, all around our wires. During extreme solar storms, "some grid systems may experience complete collapse or blackouts,"[5] according to the US National Weather Service's Space Weather Prediction Center. These storms really happen, but in modern times have luckily missed the Earth, sailing past like an especially perilous one did on July 23, 2012.[6]

Another potential cause of catastrophic collapse of the electrical grid is digital infection by malware: viruses, worms, and Trojan horses. The grid is increasingly computer controlled, a trend that makes it correspondingly susceptible to sabotage through digital infection. The risk of digital sabotage has increased since the first major such attack in the cyberwar era, the Stuxnet worm, which seriously damaged Iran's nuclear weapons program.

One can't predict the unpredictable

It is prudent to keep in mind that the lights could also go out for causes presently unknown, just as people still die of unknown causes despite all the advances of modern medical technology. The world's electric grids are highly complex systems. Unknown interactions among vulnerabilities that individually may seem manageable (to recap: grid decay, fossil fuel issues, electromagnetic disturbances, and digital infections) could turn serious problems with one into a cascade of destruction leading to overall catastrophe. It could happen; there is no proof otherwise.[7]

Fuel and food scenarios

Oil refineries are quite flammable. When attacked they easily go up in flames, self-destructing. Their size, centralization, and importance make them tempting targets for military attack. Such attacks can be overt (airplanes and bombs), or they can be covert: For example as society becomes increasingly infiltrated by computer controls, computer worms and viruses will become means for commanding machinery to go beyond design tolerances so that they fail catastrophically and are destroyed. Large, important, vulnerable, — and centralized — installations like refineries are natural targets.

As with fuel and electricity, the centralization of modern food production entails risks. Widespread disruption of either production or distribution of food would be a major catastrophe for any society based on a modern advanced economy, and centralization makes disruption more possible. Large amounts of food require large amounts of land to grow on, so the centralization problem consists of a high dependency on a sophisticated transportation network to bring food to the centers of population and socioeconomic activity. Even though crops are grown over a relatively large area, dramatic climatic changes measured in months, caused by nuclear winter, volcanic winter, or asteroid impact could lead to widespread crop failures, thereby ending civilization. Disruptions to the transportation network could be just as bad even if the crops themselves still grew green and luxuriant.

What could disable transportation of the food supply? Some risks can be identified. Most foodstuffs are delivered by truck. As anyone who has driven a large highway near a major city very early in the morning knows, armadas of large trucks can take over the highways, zooming in with daily supplies. Anything that stops the trucks would strangle the city. A deadly and highly communicable disease present in the city might scare away the truck drivers. On the other hand, a city trying to protect itself from such a disease arriving could face an impossible dilemma: getting supplies in while keeping the disease out. Fuel shortages, at least, would presumably not stop the trucks because food transportation would hopefully get sufficient priority no matter how limited fuel may be.

In a military context though, priorities might change. Civilian lives may be considered expendable, not by the wishes of the majority of course, who are civilians themselves, but by decision-makers whose priorities obviously tend toward "me and mine first," followed perhaps by military needs a distant second and civilians third (see Chap. 18). Devastating damage to roadways would halt food distribution. Blockades by roving militias could also halt food distribution. As embedded processors (the brains of computers) progressively proliferate as integral components of complex engineered systems like autonomous vehicles, computer viruses and worms could potentially be programmed to infect almost all vehicles and then destroy them without warning at, say, midnight of 9/11/2021. The Stuxnet computer worm demonstrated equipment destruction, so it is possible, unless equipment is designed specifically to prevent it.

For prevention, a simple design strategy that can help is the *hardware interlock*.[8] This is exemplified by the familiar case of an ordinary microwave oven that stops automatically (owing to a hardware interlock attached to the door) when the door is opened, thus preventing the escape of dangerous microwave radiation.

Predictions and recommendations

Where centralization creates vulnerability, decentralization alleviates it. But how is decentralization to happen? The growth of centralized electricity, fuel, and food was the result of neither an evil plot nor an ideological imperative. Rather, it was the end point of the path of least resistance, the macroeconomic result of innumerable microeconomic decisions, a rough shove from the iron fist of economist Adam Smith's invisible hand.[9] After all, it is quite impractical to build a coal fired electric generator in your closet. Constructing a nuclear reactor in your mother's shed is even less practical (yes, it's been tried).[10] Most of us can't dig an oil well in our backyards, and the few who can would find building a refinery next to it to be a poor investment decision. Similarly you can't fill a grain silo from your garden, so building one next to the vegetable garden would be financially unwise. (It would also cast unwelcome shade on the growing veggies.)

Fortunately, decentralization that does not require these things is possible. It may happen, by itself although faster, better, and more efficiently if encouraged. This is good news because it will make society more robust and our lives less susceptible to disasters. Let's see how decentralization can be our future.

Photovoltaic energy: solar cells

If only you could dig your own oil well and then build a cheap do-it-yourself refinery in your basement! Then you could pick up a home generator at the next sale at your local hardware or home supplies store and move "off grid," that is, make your own electricity and be free of dependence on the over-centralized, greedy[11] electric generation and distribution business. In fact, making your own electricity is likely to become easier over time. Home electricity production is unlikely to require fuel at all. Instead, indications are that individuals going significantly off grid will happen on a huge scale, nation- and worldwide, as a result of it simply being cheaper that way (Adam Smith's invisible hand at work, always pushing people to seek more for their money). The reason is solar energy.

Electric energy directly from sunlight, technically called photovoltaic (PV) energy, has been getting cheaper and cheaper for years. As an electronics technology comparable in important ways to computer electronics, it will continue getting cheaper until it first becomes broadly competitive with conventionally generated electricity and, later but not much later, generally out-competes it. A key event in this trajectory when "grid parity" occurs, meaning the cost per kilowatt hour of grid-supplied electric current (currently on the order of 10 cents in many places) is matched by PV. With these progressively declining prices, PV distinguishes itself from wind generators, with giant windmills and windmill farms you may hear about being a purely temporary trend. PV electricity is actually free on a minute by minute basis, since one need not pay the Sun to shine. However, solar cells, solar panels, and future solar paint have up-front purchase and installation costs, as well as finite lifetimes. Solar paint is particularly intriguing. For example, if cheap enough, people in both wealthy nations and developing nations could simply paint it on a wall or roof and draw off electricity. Solar paint containing biomolecular machinery extracted from ordinary plants as its "secret ingredient" has already been invented, proving the concept.[12] With further refinement the prospects are good for bringing electricity to even the world's poorest areas.

Figuring the cost per kilowatt hour, PV is not yet fully competitive with your local electric utility in most areas. But it's getting closer all the time.[13] When it happens — watch out! Rooftops and perhaps walls will be covered with solar panels, solar shingles, solar paint or solar siding instead of today's shingles, ordinary paint, aluminum or vinyl siding, etc. The typical house and building will be built or its roof and walls remodeled to produce electricity. Electric power will then be free when the Sun is shining. It will be sold to the grid when you have extra and purchased from the grid when the Sun is not shining. This is already done on a significant scale. Indeed centralized electricity production in the US is down from a 2008 high.[14] Not surprisingly, conventional electric utilities are fighting back, seeking to hold back the solar tide and delay the inevitable as long as possible. As early as 2013 the leading electric utility industry lobbying group, the Edison Electric Institute, called its promotion of certain solar power suppressing laws a "must-consider action" lest home solar energy become so disruptive an innovation as to "directly threaten the centralized utility model."[15]

As conversion to localized PV picks up steam, the average dwelling will be doing it, so stay tuned. I anticipate having to update this section of this book periodically as the PV revolution proceeds apace.

For those who prefer true grid independence, energy storage technologies could help even out the periods of unavailability that occur when the Sun does not shine. At the household level, that technology is the familiar battery. For example, an electric car (because it has a big battery pack) could be connected to a home electrical system. Battery technologies are improving over time though, as many a smartphone user can attest, not fast enough. Once equipped with highly localized PV technology, people will no longer be at risk of life, limb, and employment in the event of extended (or even permanent) grid blackouts.

Manufacture of electricity localized to your own home, apartment building, workplace or backyard would make individuals and society much more robust compared to the centralized approach generally used now. In an emergency, you could get electricity from your next-door neighbor almost as easily as borrowing a cup of sugar or cooking oil. Just run an extension cord.

With everything from lawnmowers to cars becoming available in electrically driven models, maybe fuel will become mostly unnecessary. But cars eat up so much energy that if you drive a fair amount you might not have enough solar electricity. Putting solar panels on the roof of the actual vehicle helps (Figure 2) but does not produce enough energy so such vehicles still need supplemental power. Also some people swear by cooking on gas stoves. So it may be comforting to know that not only electricity, but fuel could also be producible in your own home.

Home fuel production

Home gasoline production may seem unlikely given present technologies, but ethanol will become more feasible as cellulose-to-ethanol bioreactor technologies improve. Cars can potentially run quite well on ethanol-based fuel. Only inexpensive modifications to the fuel system are needed, and many vehicles are already being made with "flex-fuel" capabilities that enable them to run on either regular gasoline or the 85% ethanol fuel called E85.

Figure 2. Shuttle vehicle with roof mounted solar panel.[16]

The flex-fuel idea in cars is readily extended to permit liquid fuels and gaseous fuels to be used more or less interchangeably. The ultimate in flex-fuel engines is to replace the familiar *internal* combustion engine with an *external* combustion engine. They exist! They're called Stirling engines, after Robert Stirling who invented one back in 1816. All you need to power this engine is a heat source external to (i.e., outside of) it. Burning wood, charcoal, trash, waste oil, gasoline, ethanol, and gas all produce heat and would work fine. But for the familiar internal combustion engine, only liquid and gaseous fuels will do.

Progress toward home production of liquid ethanol is occurring, but the most mature technologies for home fuel production generate methane-rich biogas. Methane is a major component of both biogas and natural gas, which also can be used to run cars (like the Honda Civic GX, "G" for gas). If your residence has a natural gas hookup, a pump can even be attached to your gas line and used to fill a high pressure tank in your natural gas-powered car overnight. However natural gas is not suitable for highly localized energy usage, unless you are lucky enough to live over a potential gas well (*and* own the mineral rights to the land — check your title and applicable laws). In that case you could employ the services of a drilling company like the one whose highway billboards read, "Don't have a gas well? Get one!"

A more sustainable approach to gas is to generate biogas from organic waste. Methods for this have been around in one form or another for decades. Techniques are getting better and there is no reason why they should not continue to improve, until any home can have its own inexpensive biogas digester the size of a refrigerator. The secret is to decompose ("digest") organic material anaerobically. That means keeping oxygen away from it. Kitchen food waste can be an excellent biogas source. Grass clippings are another source,[17] so mowing the lawn can potentially amount to harvesting raw material for your methane digester. Like a yogurt maker, a biogas digester needs a starter with the right microbes in it; the digester should also keep temperatures and water levels right. It also will need to be able to detect when to add trace nutrients to the mix, since microbes need the right vitamins and minerals just like we do. Onsite Power Systems, Inc.[18], for example, already makes high-efficiency, two-stage digesters. The digester will need a pump and a tank to store the gas until you use it. However such systems are not small enough for a single household, at least not yet.

Going full circle, biogas can be used to generate extra electricity, and the best technology for doing that at home will likely end up being fuel cells. These turn gas into water and carbon dioxide just as burning does, but without getting hot, and they output electricity. But if electricity is the goal, it is in principle possible to get more electricity out of a given amount of garbage by going directly from garbage to electricity, cutting out the biogas middleman. A technology for that is the microbial fuel cell (MFC).[19] In an MFC, bacteria live in a biofilm on the anode and decompose the waste, which is dissolved or in tiny particles suspended in water. The bacteria digest the waste, in the process pushing electrons into the anode, from which they can be drawn off as an electric current. Although many types of waste have been found to work,[20] for small amounts of electricity, one of the most promising is human urine. At least Bill Gates, founder of Microsoft, thinks so. In 2013, the Bill and Melinda Gates Foundation awarded funds to develop urine-powered microbial fuel cells.[21]

Highly localized home food production

While electricity and fuel are pretty necessary to modern life, strictly speaking they are unnecessary to life per se. But food and water are. Highly localized water production is a solved problem: Dig a well. Or install a cistern to catch roof runoff. Water can be sterilized with cheap chemicals, such as a small amount of household bleach, or by boiling if necessary. In very dry locales, energy efficient water recyclers can work, such as the "slingshot," which is claimed to have the potential to solve water problems cheaply enough to work even in poor, underdeveloped countries.[22] Highly localized food production, however, is something of a lost art in the developed world.

At one time, a large proportion of people grew their own food, as many in underdeveloped countries still do. However, growing enough to live on, in your backyard say, is not as easy as it sounds. Try a few tomato plants and other food plants in a modest garden before cutting down the shade trees and trying to turn your whole backyard into a farm. Otherwise you will most likely just ruin the yard without affecting the grocery bill, though at least you will get healthy exercise. The "locavore" movement promotes eating locally grown food, which does not require large trucks and major highways to bring in food. Check out the nearest farmer's market, if your community sponsors one. But highly localized, home food production in quantity seems basically unfeasible: Even if you learn the skills and have a good backyard, with a regular full time job who has the time?

A solution to the home food production problem will appear sooner than one might think. It will enable growing lots of healthy organic food in one's own yard, while maintaining a normal modern life and not needing to become a full time backyard farmer. Even in congested cities without yards, anywhere the Sun shines — roofs, land strips along roadways, and even sides of buildings and windows — could be used to good effect. Such localized food production is important in part because it helps mitigate the risk of devastation to communities, nations, and potentially civilization itself if centralized infrastructure systems break down. As an important and pleasant side effect, grocery bills, consumption of doubtful food additives, weird pesticides, and bioactive hormones in bought food could all be reduced dramatically as well.

How will this seeming miracle of effortless, highly local food production happen? After all, we've already ruled out quitting our regular jobs to become urban or backyard farmers. The answer is robots. Consider for a moment the lowly Roomba, the first major commercially successful home robot vacuum cleaner (the famed UNIVAC was an early computer but did not vacuum). Successive models, and vacuuming robots from competing companies, have shown progressive improvement in negotiating rug edges and other bumps, cords, and sundry floor features while vacuuming up dirt. A natural direction for home robot evolution, then, is negotiating our outdoor floor, the ground. There is little point in vacuuming up dirt outside, but there is plenty of use waiting for a robot that can clip grass as it goes.

Mowerbots are more expensive (and bigger) than vacuumbots, but in terms of cost per mow, they are competitive with lawn care services, which therefore need to be concerned by a changing business environment. But think — a mowerbot is not just mowing, it is grass farming. For example, they could be used by sod growers, who literally are grass farmers. What we call "mowing" is, in a real sense, highly localized home grass farming. More advanced mowerbots will progressively acquire new functionality over time. Distributing fertilizer is relatively easy. Dragging water around to water a lawn just takes a heavier duty motor. Water tanks that can be docked by a robot and self-fill with water are not much more difficult to engineer than a robot docking and battery recharging system, which home vacuum robots already have. Recognizing and removing dandelion plants nestled below the programmed grass height is only the beginning of a weeding functionality, which will progress to recognizing and killing more and more undesired weed species. Diagnosing grass health issues and applying solutions is a next step. The perfect lawn, automagically.

From mowerbots to veggiebots

The home mowerbot serves as a model of cultivator robots for other crops, each with its own unique care requirements. Prototype tomatobots were invented by 2009,[23] based on a Roomba-like chassis. Manufactured by the same company (they called it the "Create" robot) it is actually cheaper than a Roomba because, instead of costly vacuuming hardware, it has an empty bin. If you wanted to grow strawberries you would get a strawberrybot. A prototype was demonstrated in 2010 in Japan. It was named IAM-BRAIN, not because it is a brain, but because it was built by the Institute of Agricultural Machinery's Bio-oriented Technology Research Advancement Institution.[24] Veggiebot commercialization got a boost in 2012 when Blue River Technology announced receiving $3.1 million in funding to develop agribots based on its prototype Lettuce

Bot,[25] which could tell weeds from lettuce with image-processing software and then kill the pesky weeds. How soon will it be before you can trade extra tomatoes, strawberries, and lettuce to your neighbor for apples or squashes grown by their applebot or squashbot? A good reason to get to know your neighbors better... making neighborliness a valuable extra benefit of veggiebot technology.

The natural next step in home veggiebot evolution after single-veggie bots is a general-purpose bot that can tend a vegetable garden with diverse kinds of plants in it, watering each as needed, weeding, diagnosing, and fixing plant health problems, squashing pest insects that are on the leaves, but leaving valuable insects like ladybugs alone, etc. People could go back to not knowing their neighbors very well, just like now, without endangering healthy and tasty meals based on a variety of crops. Ultimately, growing all the food plants that your available area can hold will be as easy as flipping the on-off switch on your veggiebot. By then, the dangers of centralization in food production will have been significantly mitigated by highly localized, labor-free food production, using robots.

Chapter Seven

When Genomes Get Cheap

We are acquiring ever vaster amounts of information on how variations in our 20,000–25,000 genes affect our physical, mental, and medical characteristics. This will have major impacts, from medicine to self-knowledge.

Your genome is unique, unless you have an identical twin. Most cells in your body contain a copy, red blood cells being a notable exception. The cost to "sequence" your unique genome (techspeak for figuring out what it says) is decreasing fast. The US government spent $2.7 **b**illion to sequence the first human genome, finishing in 2003.[1] James Watson, co-winner of the Nobel Prize for co-discovering the structure of DNA, had his genome sequenced for $2 **m**illion in 2007 (that's a price reduction of over 99.9%).[2] A dramatic rate of decrease has continued. With genome sequencing costs breaking the $1,000 barrier in 2014 according to Illumina, Inc. (exactly how the cost should be calculated is a matter of debate), your genome will likely be sequenced as part of your medical record in the near future. Soon it will be possible for nearly any disease, condition or characteristic you can imagine to be checked to see how much it is influenced by each of the roughly 20,000–25,000 human genes.

Ultimately we'll know almost anything you could think of about how a person's characteristics are associated with variations in each gene and its related molecular machinery that regulates when and how much the gene is activated. These characteristics will concern personality (a tendency toward cheerfulness or melancholy, Myers-Briggs test results, faithfulness or a tendency to cheat, ability to delay gratification, risk aversion, and on and on); physical traits ranging from eye color to height and weight; intelligences from verbal to mathematical, musical to artistic; and disease susceptibilities from flu to cancers.

Organizations and clubs will doubtless spring up for people with specific gene variations. A cultural movement including digital communities as well as traditional in-person clubs and social organizations will probably arise based on one's genes, particularly genes that are discovered to have personal meaning. Good candidates for that include genes related to personality characteristics, susceptibility to various disease conditions (take your pick), and variations in dietary needs. Awareness of diet-related genes will probably enable at least some people to feel (and be) much healthier than they otherwise would. Awareness of one's zodiac sign would, logically, be replaced by awareness of one's appropriate genes ("I'm a Rs6265 Met/Met" … "Interesting, I'm a Met/Met too!"). Astrologers will have to find another line of work; perhaps they will rebrand themselves as "genologers."

Many personal characteristics are impacted to small degrees by variations in numerous genes. After all, any gene whose variants have a noticeable effect will also slightly affect the tendency to worry, because people can worry about almost anything, such as those effects. That does not automatically make those genes "worry genes"! Such factors make figuring out what genes do more complicated, but researchers will be undeterred. There is need and opportunity for huge amounts of research on why variations in this or that gene and its regulation affects this or that characteristic. Schizophrenia is one example that is already attracting this type of attention. Pick whatever personal characteristic comes to mind, and that's another example. This wave of research will last a long time, because checking the connection of each of 20,000 genes with each of, say, 10,000 personal characteristics means up to $20,000 \times 10,000$ possible things to check. That's a lot — 200 million! Synergies and interactions among genes and their variants will be another area demanding extensive research. Adding yet another layer of challenge, people will also want to understand the roles of complex interactions among environmental variations and gene variations, gene methylation, and protein variations arising from differences in what pieces (exons) of the corresponding genes are transcribed.

Health problems, whether physical or mental, serious or minor, will often be treated in light of the sufferer's genome. When leukemia researcher Lukas Wartman came down with leukemia himself in 2011 his leukemia researcher colleagues hit the ground running, using full-genome analysis to rescue him from what appeared to be certain death. Most of us don't have colleagues like that. Yet, "[w]ith a steep drop in the costs of sequencing and an explosion of research on genes, medical experts expect that genetic analyses of cancers will become routine."[3] This is already routine at, for example, the Myeloma Institute for Research and Therapy. Then cancers will be treated in the average sufferer based on the specific genetic mutations that made the tumor cells cancerous. For other illnesses as well, drug doses will be optimized genomically because people have greatly varying sensitivities to medicines. Do you have a prescription now? Your dosage is probably not genetically optimal. That will change. In fact, almost any medical recommendation could potentially be tailored to the sufferer's genome.

Some gene variations are rare, so such a variation may cause a rare disease. An orphan disease is one that is rare enough that the pharmaceutical industry does not try to develop medicines for it (they won't "adopt" it, so it's an "orphan"). Officially, an orphan disease in the US has anywhere from 1 up to 200,000 sufferers domestically.[4] Medically, orphan genetic diseases will be tracked down with increasing ease, though not necessarily cured. Thousands of orphan diseases are known, some more serious than others.

As just one example, Lafora syndrome is a genetic disease affecting about one person in a million. Patients have normal childhoods, but typically become ill during adolescence with symptoms such as seizures, hallucinations, decreasing intelligence, ataxia (poor coordination), progressive dementia, and death. At the same time, Lafora bodies — abnormal lumps of starch-like polyglucosans — build up inside the victim's cells. A drug called Zonisamide, more commonly used to treat epilepsy, helps some. Luckily, if you are reading this you are probably old enough to be home free, and can quit worrying — at least about that.

Game changing genes

When large numbers of personal genomes become available, the race will be on to find rare gene variations, many of which do not cause rare diseases. Rare variations will enable us to understand the limits of human nature — the most unusual among us, genetically speaking. A few of us may be evolutionary "game changers." Consider for example the game changer that makes us fundamentally different from our closest animal relatives, the chimpanzees and bonobos ("buh-NO-bos"). We all have teeth, four limbs, eight fingers (+ two thumbs), fairly big brains, and so on. What makes us so unique, even more than the difference in brain size, is our language skills, which are incredibly advanced if not 100% fauult-free.

The human version of the FOXP2 gene was among the first to be identified as necessary for our language abilities.[5] Although FOXP2 is far from the full story behind human language, it exemplifies the concept of game changer genes. Surely, in the many billions of human beings currently alive there must be some potential game changers! It is tautological that game changers have made us what we are, and the ancient populations that produced the gene variants that culminated in modern humans were much, much smaller than today's several billion.

What will we do when we find that lucky (or maybe unlucky) person with a new, game changing gene? Clone? Kill? Instead, expect at least some future prospective parents to splice the new gene into their sperm cells right at home, perhaps after purchasing a Model 2100 Improv-Ur-Sperm device. (Batteries not included. Sperm-collecting attachment sold separately. Plastic turkey baster for easy, convenient at-home insemination included FREE — and as an added bonus may also be used to baste turkeys.) This is not so far off; "CRISPR"-based genome editing was used to edit the DNA of mouse sperm cells in 2015[6]. The CRISPR approach is so powerful that in March, 2015, scientists proposed banning its use on humans.[7] But you can't hold back the tide for long. Case in point: The very next month, scientists in China used it experimentally to genetically modify fertilized human eggs.[8] Actual genetically modified people could be as little as 10 years away.[9]

Ultimately, superbabies will be designed and raised, whether with game changing genes or merely with desired combinations of normal genes.[10] Already, a technique called PGD, for preimplantation genetic diagnosis, is being used for gender selection.[11] Even without help from CRISPR, PGD could eventually be used for selecting any genetic characteristic for your baby. Here's how it works. Mom undergoes a procedure to harvest a bunch of egg cells, which are then fertilized with dad's sperm, and the resultant embryos grown in the lab until they each have eight cells. One cell in each is removed and genetically tested. Then one embryo with its seven remaining cells is selected from the bunch based on its genetics and implanted into the mother's womb. There it compensates for the missing cell and eventually develops into a baby. The process is not very romantic, but that's ok because many people will do anything for their kids.

Once the limits of human variation are mapped and explored, hopeful parents and evil dictators will take keen interest. Desired genes might be considered national resources to be propagated and increased. This might speed evolution of the species. Alternatively, authoritarian governments may suppress genes that lead to people with the capacity to threaten them. Such autocratic soldiers of Satan have historically viewed successful people as a threat and sought to neutralize them because of their leadership potential.

What to do

You can already get various genetic analyses done by companies that specialize in testing samples by mail. At this writing, 23andMe will do it for $99, telling you what population groups your ancestors came from. That is just an example, and competing companies exist. If you are (or plan to be) an ambitious young scientist, consider a research interest in mining the linkages between genes and phenotypes (physical characteristics) from large sets of genomes. If you have executive authority in a genetic analysis or other biotech company, there is still time to take advantage of commercial opportunities in this area. Angel investors and venture capitalists might consider how to avoid being left behind in this area. Research funding agency program directors can be alert to this research trend.

If you are none of the above, consider having a small tissue sample frozen and stored for your children and grandchildren to analyze. They may need to sequence your genome to better understand theirs, enhance their own health, and treat their own medical conditions.

Chapter Eight

Cheaper Teaching, Faster Learning

The growth of online learning means that one instructor could oversee a course with a million students in it. What are the economic incentives involved and what will this mean for education?

Today's students are mostly still educated like they were over 3,000 years ago, when the first records of classes taught by an instructor appear.[1] That was approximately 2040 BCE, in Egypt during the reign of Mentuhotep II, the 11th dynasty pharaoh. The tomb of Kheti, his treasurer, calls upon graduates to behave well.[2]

No word on whether Kheti wanted the graduates to apply what they were taught, or was just hoping they would act better than they did in class. However it seems there were classes composed of students taught by an instructor, just like today.

But now things are changing. Distance education is increasing because basic economic forces are pushing it along. The traditional paradigm of class, classroom, teacher, and schedule has both higher cost and lower convenience than distance education. Thus more distance education will save both time and money, making it a path of least resistance incentivized by the lure of faster and cheaper. The "invisible hand" of Adam Smith, father of the economics field, is busily at work as you read this. It's hard to fight that off forever, so a sea-change in education seems inevitable.

There is yet another way that Smith's invisible hand drives these future changes to education. The tradition of groups of students meeting in classrooms requires people to spend time and money traveling daily or even moving to another city, as well as dealing with scheduling difficulties. Institutions must build and maintain classroom space, parking lots, and so on. Online learning sidesteps those requirements, saving significant student, teacher, and institutional resources. So, how could online education *not* continue to grow as the technology available continues to improve? Eventually, it seems likely to become the rule rather than the exception. The vast potential benefits incentivize it to happen.

Duplication of effort

Looked at more closely, the issue is really a connected web of subissues. An important one is duplication of effort. There is an immense amount of it, and it is extraordinarily expensive. If there are a million students with a student-teacher ratio of 20-to-1, that's 50,000 teachers. Teaching is a socially valuable profession, though less honored and lower paying than most professions. More on that later, but still, the cost of employing 50,000 instructors adds up. Suppose the cost per teacher was around $100,000 a year (including not only salary, health care, retirement contributions and other benefits, but also support personnel from expensive administrators of endless variety to secretaries and janitors. That is... let's see... $5 billion per year.

Now suppose each student learned at their own artificially intelligent computer, with one teacher overseeing each subject for all million students. Each lecture (and potentially, supplementary and explanatory links and material as well) would be accessed by each student, a million times overall. Each homework and quiz would be scored by the computer. Every wrong answer would be linked to a short (or maybe long) supplemental lecture also provided by that one teacher. Each student could proceed at his or her own pace. The one actual live teacher would spend much of the time analyzing currents of misunderstanding and other holdups, and recording mini-lectures and course updates to fix those misunderstandings and otherwise improve the course. Automated help desk systems would answer most student questions, and online forums would enable other students to answer most of the rest.

While a million students in the same class has not happened yet, Stanford University offered "Introduction to Artificial Intelligence" online in 2011 to 160,000 students. With that success the professor founded startup company Udacity to offer online courses to huge numbers of students at a time.[3] While one pundit claimed graduate degrees might end up costing $100, Sebastian Thrun, the Udacity founder as well as a well-known roboticist, stated baldly, "It's pretty obvious that degrees will go away" since "this model isn't valid anymore."[4] By 2015 he had toned down the rhetoric just a bit, saying he now saw Udacity as becoming the "Ikea" of education.[5] Interestingly, his colleagues Daphne Koller and Andrew Ng, like Thrun in Stanford's Dept. of Computer Science, also started their own competing online education company, Coursera. The three must have started out seeing eye to eye, but ending up cross-eyed.[6] Adding to the confusion, in Summer 2013 Stanford itself adopted a 3rd competing MOOC, OpenEdX.[7]

Distance education cannot easily substitute for teachers whose subjects are not realistically teachable at a distance. Kindergarten and primary school teachers are examples. In higher education, subjects like music and theater, for example, need traditional instruction more than other subjects more easily taught using distance education technologies. Some people cite technical subjects as good candidates, though the influential and selective Olin College of Engineering is the focal point of a movement toward interactive, face-to-face, small-group technical education. One might think an ideal topic for a huge online course would be how to teach huge online courses — but an early offering of such a course became a notorious flop.[8] Despite the problems, the lure persists.

If today's method of thousands of teachers teaching the same things sounds like a lot of duplication, while having one teacher doing all of it instead sounds efficient, consider this: In both cases the number of students learning the same thing is many times greater than those thousands of teachers! Imagine a hyper-efficient world in which labor specialization and its associated educational requirements are taken to the logical extreme, so that any specific topic is learned and later used by just one student. If that stretches the non-duplication idea past the breaking point, it is undeniable that students generally learn some things they don't really need to know. This is extraordinarily wasteful: If a person starts kindergarten at 5 years of age, graduates from college and starts work at 22, and retires at 65, that is 60 years, of which over one quarter is spent in school. While child labor is thankfully a thing of the past in civilized countries (where "civilized" means, in part, no child labor), it is not a requirement to wait until 22 to work. Even for someone who does wait, replacing study of less needed topics with study of more important ones would result in a more efficiently trained 22-year-old.

Unfocused curricula — good or bad?

Higher education can seem bedeviled by lack of focus. Four year degrees often take more than four years, particularly for engineering majors who still, all too often, graduate never having learned to communicate effectively. Two-year degrees have a similar problem: Two years is often longer than is actually needed to enable the career advancement that motivates the pursuit of such a degree. Certificate programs are shorter, more focused, and hence more efficient. Sometimes a 4-year degree plus a certificate is a great combination for employability — disproportionately due to the certificate. As with the duplication issue, this lack of focus in traditional degree programs is inefficient, which is to say unnecessarily expensive in time and money. Hence change is incentivized by economist Adam Smith's ever-present invisible hand.

Changes to expect include more demand for certificates that are based on just a few college courses and a consequent decrease in societal focus on multi-year degree curricula; students taking a single course to meet a specific educational goal, where the course title, syllabus, and grade constitute a certificate of mastery of a specific topic; and students achieving a collection of certificates and useful individual courses in addition to, as part of, or instead of a single four-year degree. Additional changes that may occur are attempts by colleges and universities to sell education by incentivizing individual professors to promote enrollment in their courses, because more enrollment in a course means more income from it. This would work better in an educational environment rich in certificate-seeking students, because such an environment will have fewer required courses compared to the number typically required for four year degrees. Required courses tend to have inflated enrollments compared to what the enrollment would be if students were free to enroll or not based on their perception of the value of the course. If you have been to college you can doubtless appreciate the magnitude of change in the college experience that would ensue if students pursued collections of certificates instead of single multi-year degrees.

Advantages of traditional higher education

If the future of higher education is one of distance ed, computerized efficiency, and certifications in specific areas, what about the future of traditional education? That is, what about intellectually broadening, reflective education that enhances critical thinking skills? That tradition arose from many centuries of advanced education aimed mainly at enabling an elite few to become leaders of business, military, and government. The focus in modern times on higher education for as many people as possible, incentivized by the needs of modern economies, correspondingly *dis*incentivizes providing the more traditional, broadening type of education to the general public. Here are the main disincentives.

1) Some powerful institutions have an interest in indoctrination — getting people to think in the way the institutions want. But enhancement of critical thinking skills conflicts with indoctrination. Indoctrinators benefit from *weakening* peoples' thinking skills.

2) Awareness and wisdom-enhancing education competes for students' time, money and attention with study of more focused, specific skills.

3) The powerful (like everyone), have an incentive to maximize their own interests. Promoting those interests in a democracy is sometimes most effectively pursued by cloaking them, because implementing those interests requires the support of the general public who, if they better understood the situation, would instead support their own, different interests. Exemplifying a perennial hot-button topic, consider taxes. Progressive (but not regressive) taxes recirculate money that was harvested by the wealthiest from their employees, customers, and society generally. Many people would approve of this recirculation function of taxes because it redistributes the national wealth to benefit more people. To undermine this approval, highly funded political machines distract and redirect attention using epithets like "job-killing" and the like. The effectiveness of this cloak of distraction depends in part on the general public failing to see behind it. But education enhances wisdom, awareness, and ability to read between the lines, enabling people to better see through clever political propaganda. This fact about education is an incentive to disparage education, educators, and educational institutions by any lobbying group that wants to shrivel the kind of

education that leads to wisdom and awareness in the general public in order to better promote its own agenda. (Full disclosure: I am an educator.)

An enlightened and aware citizenry can better defend itself from the brainwashing manipulations of those seeking to control what we think. Learning such basic defensive skills need not depend solely on a college education, which in any case fails to help those who do not go to college. So why not start in grade school? As a second grader I struggled through an English exercise that required classifying statements: Are they *fact* or *opinion*? It was tough slogging for a second grader but, unlike nearly everything else in second grade, it made enough of an impression that I still remember it. Let us expand that concept by also teaching people to identify the statement writer's *agenda* and thus any incentive to manipulate the reader's mind.[9] This idea is far from new; applying it more will reduce brainwashing.

Longer term

Many things that need to be learned now won't need to be in the future. The past provides examples. Calculators are so ubiquitous that learning to calculate by hand is less crucial than it once was. How often do you do long division? Do you even remember how? Ever more advanced calculating devices do not make math courses obsolete, but do change what is taught. Instruction will focus more on defining problems and less on solving them — since computers will do that, once problems are defined. Cursive handwriting is no longer taught extensively, because of keyboards, which themselves are already, bit by bit, giving way to speech recognition. Spelling will soon be taught less, owing to the general availability of spell checkers and the inroads of speech recognition. This frees up study time that will be spent studying other things instead. This trend will only continue. What is next? Grammar will become increasingly unnecessary to study as automatic grammar checkers continue to improve. Learning to read a foreign language is getting less important now that web search engines are providing translations online, in real time, whenever needed. Speaking is next. Real-time speech translation will make it possible to travel anywhere and talk to anyone, regardless of language, with just an app that hears and translates instantly. Uncounted millions of students worldwide will no longer need to spend uncounted hours learning English, and myriads of English-speaking students will spend less time on other languages so that they can spend more time on... what?

As computer interfaces improve, academic subjects may be available on small devices built into glasses. Google Glass was the first major glasses product. However they retrenched in January, 2015, while excitement focused on Oculus Rift. Regardless, with further progress the earpieces of such glasses — or small hearing aid-like devices instead — will sit behind your ear and talk to you. Call them eartop computers.[10] The eartop's ancestral roots are in Apple's iPhone 4S, "S" for Siri, the name of the first major personal assistant with speech interaction for pocket electronic devices. Siri was born (er, released) on October 14, 2011. Such eartop devices will be much better than Siri. For example they will contain entire textbooks on whatever subject you need to know about (or a dozen textbooks, or thousands) and will be able to converse with you about the material using QA algorithms (you ask it **Q**uestions and it gives **A**nswers). Facts will therefore become less important to commit to memory. Such devices will hear and understand what you say. Eventually they may read your brain waves directly, by-passing your vocal chords.

While communicating back to you by audio using an earphone or bone conduction is a reasonable approach now, sending messages into your brain via electrical, magnetic, or ultrasound signals seems likely to eventually work too. Such signals may also be useful for other purposes, like altering moods, or turning on a desire to study. If these technologies develop to the point of working significantly better than drinking coffee and such, the incentives to use them in conjunction with their potential long-term effects on the psyche will become quite worrisome. There is a fine line between a helpful nudge in the right direction and destroying yourself as effectively as any present-day drug user. Think of science fiction author Larry Niven's "wireheads" as a scary exemplar of the possibilities.[11]

Far more than just tools, tiny eartop computers will become more like guardian angels — or at least ever-present personal assistants — always there to augment reality as personal knowledge sources, advice givers, and even companions.

At this point, teachers will be unnecessary. Of mainly historical interest, people (and their eartop guardian angels) will wonder at the old days when the phenomenon once called "schools" existed. Teaching will have been replaced by computing. A human without an eartop guardian angel will be as handicapped relative to other humans as a human today who is deaf, blind, and mute. Education and much else will be a thing of the past, and life will be hard to imagine but very different.

What should we do?

To many observers, change seems inevitable in higher education. For decades, a disruptive shift toward distance education has been touted as just around the corner. This time — maybe — it really is. To smooth adjustment to the shift, there are two major response strategies for higher education that society can demand and public institutions of higher education can provide. The first is for universities to develop unique angles in what is taught and how, so that they are not providing carbon copies of each other's classes. This can mean seeking niches in which to provide unique degree programs. It can also mean teaching standard courses, but in unique ways that can attract remote students from far and wide who need to learn the material and want it packaged in a unique way that suits their background, learning style, etc. A clearinghouse website would be needed to match students with courses, sort of an online course bazaar. Higher education would need to develop ways to permit students to take online courses from numerous different providers and count them all toward a single degree, certificate or other curricular package.

The second response strategy is for the United States in particular to eliminate the artificiality of how public universities tend to protect their cookie-cutter, carbon-copy classes and programs now: Tuition rates are typically much lower for students in the same state than the higher, often private college-level rates charged to students who are residents of another state. Many countries are way ahead of the US on this issue. The problem is that this protects universities from having to compete for students with universities in other states. Even though a considerably more desirable university in another state may offer the same programs, students will still go to a university in their own state because the fees to attend an out-of-state university are so much higher. The process works alright for students in states with top universities, but students in other states are locked out of the chance to go to those universities by the price differential. This artificial price structure ensures that public universities do not have to compete with universities in other states. If they did have to compete, some might not be able to, but overall quality of the educational experience, on average, should go up.

Artificial, protectionistic tuition barriers in the US and wherever else they exist should be eliminated. The rise of long-distance education may encourage that. Universities would then need to get much more serious about competing. They would improve their own programs to match those of the best, or branch out by designing distinctive programs offered at few other places and therefore at which they can more readily be among the best. The educational process would work better and the careers of graduates would then work better as well. Society would win; it should be done.

Chapter Nine

Soylent Spring

Artificial meat can in principle be cheaper than animal meat. So, when we finally figure out how to make it taste as good as ordinary meat but at a lower price, society will switch in a big way. There will be no going back.

The movie *Soylent Green* explores a dystopian future in which most people eat "soylent," processed wafers of different colors. The best tasting, most sought-after of these is Soylent Green, and it's in short supply. Detective Thorn sneaks into the factory where it is made, discovers the closely guarded secret ingredient, and is marked for death. Later, the injured Thorn seeks to spread the word in the movie's perhaps most memorable line, shouting "Soylent Green is *people!*" because it is manufactured from — human corpses.

Yuck. Hopefully, this will not come to pass. We are, however, in the dewy springtime of the future history of human food. Though our diet now is different from what it once was, these differences are just a taste of the likely changes ahead.

Some History

Our hunter-gatherer forebears had a varied and generally healthy diet of meats and vegetables, grown strictly organically by Mother Nature. Then agriculture was invented, most famously in the fertile crescent, which was centered around Mesopotamia in what is now Iraq. That was perhaps 11,000 years ago. Later, agriculture was reinvented in China, Africa, and the Americas. But the earliest may have been New Guinea, with intriguing evidence suggesting taro and yam farming over 12,000 years ago. The nearby Solomon Islands may have preceded even that, with some evidence of taro cultivation well over 20,000 years ago.[1]

While a diet top-heavy in any particular farmed plant is not as healthy as a more varied hunter-gatherer diet, farming can support a much higher population density. Thus humankind's first major dietary shift was, from a health standpoint, in the wrong direction. Call it the first step toward a diet based on potato chips and soda pop if you like, a goal whose full attainment we may hope remains forever out of reach.

The industrial revolution enabled further significant changes to agriculture based diets. On the one hand, a much greater variety of foods are now available to the average person in a developed country than was the case earlier. This is healthy. On the other hand, large amounts of highly processed, nutritionally deficient foods are now widely consumed (like potato chips and soda pop), as well as foods containing a wide variety of chemicals not normally found in nature. These range from food additives not always proven safe, to chemicals like pesticides in plants and antibiotics in animals, to other clearly noxious substances like the small quantities of arsenic•often fed (deliberately!) to chickens.[2]

That is the history of food in a nutshell and brings us to the present day.

What's on the menu next?

Meat. More specifically, artificial meat. Humans are built to like meat, but artificial meat is well positioned to be a disruptive technology. Not because so much ordinary meat is laden with unhealthy levels of cholesterol-enhancing fats, even though it is. Nor is it because the most efficient meat animal production requires facilities, called factory farms, that are more like factories than farms and are inhumane to their animal inhabitants — even though they are. It's not because cattle fart into the atmosphere about 80 million tons (73 million tonnes) per year of the potent greenhouse gas methane,[3] even though they do. Nor is it because producing one pound of animal meat uses about ten pounds of animal food, the growing of which competes with growing human food, making human food scarcer and more expensive and reducing the human population that the Earth can support — even though it does. While those are all serious problems, the world shows no sign yet of a major shift from animal meat to artificial meat in order to solve them. So what might cause such a shift?

Taste and cost

A lot of artificial meat products taste neither like regular meat, nor as good. Since people eat meat mostly because they like the taste, until the taste issue is solved the artificial meat industry will continue to operate with one hand tied behind its back. However as advances in artificial meat continue, the taste problem will become history. Animal tissue grown in a vat instead of in an animal is one approach. PETA (People for the Ethical Treatment of Animals) offered a $1 million prize in 2008 for a commercially viable process.[4] Although it went unclaimed, on August 5, 2013, the first hamburger made from vat-grown meat was fried up and gulped down, at the bargain basement price of only 250,000 Euros (about 330,000 US dollars).[5] The intent is for the technology to improve, leading to lower prices and commercial viability. Anticipating that, *The In Vitro Meat Cookbook* appeared in 2014.[6] Meanwhile other strategies are already on the market.

Quorn, for example, tastes pretty good, but is expensive. Made from mycelium (fungus roots — yum!) it tastes surprisingly like chicken. But that's not the only product. Others are getting good too as the technologies improve. One such product that came on the market in 2012 was, according to one reviewer, "so good it will freak you out."[7] The company plans to do what any self-respecting artificial meat company should aspire to — sell it in grocery store meat departments. With the taste problem essentially solved, the remaining obstacle is cost. Delicious artificial meat is generally more expensive than animal meat. Until that changes, artificial meat will remain a niche specialty.

Making Artificial Meat Cheap

In principle, artificial meat has the potential to be much cheaper than animal meat. The reason is the intrinsic wastefulness of growing meat animals. With about ten pounds of feed needed to grow one pound of meat, it would be about ten times cheaper to convert feed into artificial meat pound for pound, if someone would only invent the technology to do the conversion. While such a 1:1 conversion ratio may be impossible, the present, the wasteful 10:1 current rate certainly presents an opportunity for vast improvement. As the technology advances it is hard to believe the cost issue will remain unsolved. Long term, even growing plants is a wastefully inefficient way of using solar energy to create food. Eventually food production (including artificial meat) may be done by chemical synthesis of nutrients. This is a chemical engineering problem, was recognized as such as early as the 1890s,[8] and is the logical endpoint of the soylent concept.

Watch out meat animals, you're going obsolete!

Once the taste and cost problems are finally solved, the meat industry is in for a major shock. Shoppers will be free to think, "Why buy animal meat when artificial meat is as good — or better — and cheaper, too?" Not to mention healthier, more humane, ecologically beneficial, and even good for other food prices as well as world hunger.

Recommendations

If you are in the meat industry, start thinking about positioning yourself for a career shift. If you eat meat, get ready to save money. Where's the beef? It's being made in the lab. Got artificial meat? It'll be the real thing. Real people? They'll be chowing down on real food — artificial meat. It's what's for dinner!

Chapter Ten

The Turbulence of Short-Term Action

Society is stuck in a short-term mindset, but we shouldn't be. Oddly enough, picking the best actions of an organization or society with respect to a far off end point provides better average yearly performance than optimizing each year individually, one at a time.

Remember the story of *The Ant and the Grasshopper*?[1] The ant toils tirelessly, preparing for the future. Meanwhile the grasshopper lives it up, ever the party animal (literally). But when winter arrives, the ant is prepared while the grasshopper starves. This enduring tale shows our cultural understanding that short-term action has risks. Risks run from personal to global, at time scales from seconds to the distant future. In fact, the poorer prospects of short-term optimization compared to longer-term optimization is a provable, logical fact, as we will see.

Let's begin with personal risks on a time scale of seconds. Did you run a yellow light today? Go through a green without checking cross traffic? Turn in front of an oncoming vehicle? How close, how fast, and with how much general congestion? Such actions assume they can be completed without something unpleasant and unanticipated happening within the following second. That's predicting the short-term future, and it has risks because the future is, to a significant degree, unpredictable.

Moving to a longer time scale, how about hanging out instead of studying an important topic in a college class? Sure, many assignments truly are unimportant. Moreover, grades don't say that much about future success. However for the right topic, an hour of learning is probably statistically worth cash in terms of starting salary and lifetime earnings. Writing, technical skill development (depending on major), and so on are among possibly valuable topics. Similarly, if you smoke, each cigarette shortens life by an average of 11 minutes.[2] If you don't smoke, trying it risks becoming a long-term smoker because nicotine infiltrates the brain like a dangerous secret agent. These examples illustrate that short-term action can cause long-term risks.

As another example, saving for your retirement makes sense. But all too often, short-term action wins out and money is spent that would have been smarter to put into a retirement account. To compensate, the US and many other industrialized countries have created systems to encourage retirement saving, imperfect though they may be.

Shifting our attention from individuals to groups, companies are groups where many people spend much of their waking time. Companies are notorious for short-term action. Many are more interested in maximizing performance over the next year than over the next 10 years. But sometimes maximizing 10-year gain requires some degree of sacrifice for a year or two, analogous to an individual tolerating a low income while getting an education in order to improve chances for a lifetime of higher income. A short-term maximization strategy will not make that sacrifice. The likely consequence, paradoxically, is a poorer average yearly result than if 10-year gain was maximized from the start. Why?

Logical argument

Maximizing with a 1-year horizon, year after year, cannot in principle be expected to do better over 10 years than spending 10 years focused on an end point that starts 10 years out and creeps one year closer with each passing year. Why? Because the 10-year result was maximized intentionally in the latter case, but in the former is merely the accidental result of 10 successive efforts aimed at 10 different end points. An intentional 10-year strategy will probably do better than an accidental one, provided the intention is implemented rationally. Moreover, a better 10-year result provides better year-by-year results, on average, than a less good 10-year result. Thus the best way to maximize the average *yearly* result is to focus from the beginning on maximizing the *10-year* result, rather than on 10 successive 1-year results. To make this into a general rule, if maximization is done rationally then a longer-term horizon will tend to produce better average short-term results than a corresponding sequence of successive short-term horizons. If that's the silver lining, the cloud is the higher risk that any given short-term result will be worse than if it had been specifically optimized.

Governments

The same tendencies found in individuals and groups are found in national governments, which are in fact largish groups. A classic example of short-term action is British Prime Minister Neville Chamberlain's proclamation of "peace in our time" on September 30, 1938, followed less than a year later by a declaration of war on Germany. Similarly, politicians in democratic societies regularly manage their country's affairs with their eyes glued to the next election. Longer-term national interests are thus de-emphasized, hardly a desirable tendency. Politicians in non-democratic countries are subject to lesser such forces. Of course those countries have other pitfalls — as Winston Churchill famously said in 1947, "… democracy is the worst form of government except all those other forms that have been tried …."

The ultimate result is suboptimal leadership, which can only hurt. Short-term action undoubtedly explains in part an important historical phenomenon that humanity has been condemned to repeat with annoying regularity: the parade of nations that achieve major world power status only to lose it. From Genghis Khan to the Roman Empire to the Ming dynasty, Portugal to Spain to the twelve-year eyeblink of Germany's thousand-year reich, Great Britain to Mother Russia etc. If the cycle is allowed to continue, what powerful nation will be next? The US?

Humans now live in a global village. Events in one location can affect other locations thousands of miles away. For example climatologists have determined that CO_2 emissions at your location contribute to global warming. Risks from potential pandemics like bird flu, as well as existential risks like asteroid impact, nuclear war, and nuclear or supervolcanic winter are of concern to everyone everywhere — or should be. Short-term action about such issues increases long-term risks and costs. Thus the current cost of CO_2 emission now is very low: Just squirt it out the smokestack and the exhaust pipe. But the future costs are higher and will affect future generations.

True cost is, logically, some composite of current and future costs. Thus the rational approach is to understand the true cost, then determine how to save on true costs by preemptive action. As long as true cost is lowered by more than the cost of preemptive action, the world is better off with the preemptive action: Less pain now can be better than more later. Yet short-term thinkers naturally focus like a laser beam on the "pain now" part of the equation, ignoring the all-important "more pain later" part. Long-term action requires understanding the difference.

Why do people risk short-term action?

Acting short term while ignoring long term consequences seems like a bad idea, yet it persists. Table 2 gives some reasons why.

Reason	Example	Comment
Unwillingness to delay gratification	Procrastinating, while unpleasant, seems better than tackling something anxiety-producing head-on.	Leading to the next row…
Slippery slope obscures longer-term effect	One procrastination event may be harmless but, added up, they risk poorer results. Another example: Every nuclear weapon made deters enemy attack (good), but leads toward threat of MAD (mutually assured destruction — bad).	People know about slippery slopes but procrastinate anyway, leading to the next row…
Urges and impulses of non-rational origin	You want to not procrastinate, but procrastinate anyway.	Adolescents can act on impulse as though they think they are immortal.[3] Adults often think that souls really make them immortal,[4] perhaps because it is logically impossible to imagine what not existing feels like.
Poor incentive structuring	In the 2008 financial meltdown, US bailout billions boosted big bank boss bonuses.	The bailout was pushed through by claiming an emergency, exemplifying the manipulative value and chaotic results of shouting "fire!" in a crowded theater.
Failure to understand long-term consequences	The former Easter Island civilization experienced destruction of all its palm trees, leaving no way to build watertight boats.[5]	What could the rat that ate the last palm seed have possibly been thinking? Modern civilization faces analogous situations.
Short-term fluctuation obscures long-term trends	Long-term global warming tends to lose urgency for every year that fails to set a heat record.	Fluctuations make underlying trends harder to see and to care about (also called "creeping normalcy").
Urgent fire-fighting requires ignoring longer-term risks	In a famine, people will eat the seed grain.	With no seed left to plant for next season's harvest, the famine worsens.

Table 2. Some cases of short-term action.

What can we do?

Go ahead and act short-term, but only when appropriate. For example, a salesperson in the midst of working on the next sale should focus on the sale (and its commission) without being distracted by concerns that the company might go under along with the sales position. Such concerns are valid, but for later. Focusing on the short-term goal of making a commission on a sale without distraction is a useful step in the long-term goal of earning a living. Ergo, short-term action sometimes helps produce the best long-term result. But not always...

Know when to act long term

Suppose you are driving to someone's house located 5 miles northwest. Acting short term, you soon find a road going due northwest and take it, only to discover that it curves around and leaves you going back in the opposite direction, southeast. Luckily you soon find another road going approximately northwest. Taking it, you soon reach a T intersection and must turn right or left. Though neither way points toward your destination, going right seems a few degrees better than going left. Traveling miles out of your way before finally getting to a bridge over the river, you finally make your way to your destination without too much further trouble. Having learned your lesson, next time you consult a map, go online for directions, or use your GPS app. That enables you to avoid turns that seem good at the moment, but are actually bad. So you avoid last time's wild goose chase by taking another bridge that you missed the first time but is in fact a much better choice.

Whether you are traveling, picking next semester's college classes, taking the next step in your career, doing selective breeding for crop improvement, or engaged in Middle East shuttle diplomacy, the principle is the same: Compared to short-range planning, longer-range planning tends to zigzag less, leading to better results quicker. That's why it makes sense to save for retirement, vote for politicians with thoughtful vision instead of merely talent for manipulating the emotions, and have your own personal strategic plan. It is why one of "The 7 habits of highly effective people"[6] is to "Begin with the end in mind."[7]

People care about their and their children's futures. Do they care less about their grandchildren's futures? Great-grandchildren's futures? What about ten, a hundred, or a hundred thousand generations hence? How much should we care about our descendants 10 million years from now who, if they exist, may resemble us no more than we resemble the chimpanzees, from which we most likely split a comparatively brief 5–7 million years ago? Some things to think about.

How to think and act long term

When people habitually account for longer-term horizons in their actions, long-term social goals will have a greater role in public discourse. Then, the future of nations as well as humanity will be safer and wealthier. But this may be trickier than it seems. The further into the future one considers, the more uncertainty applies: Peoples' careers take unexpected turns; national economies ignore "expert" predictions; the world is repeatedly blind-sided by unanticipated conflicts, and so on.

The best-laid plans can go awry. Next-day weather forecasts are usually not bad, jokes aside, but 10-day forecasts are more iffy; 20-day forecasts can be little more than statements of historical averages. Uncertainty expands the further ahead one tries to predict — so "eat dessert first."[8] The classical approach (and you can do it yourself with pencil and paper) is to list the possible futures, the value and probability of each, and calculate each future possibility's "expected value" as its value times its probability. Getting $1 million with a 10% chance thus has a true value of $100,000. Decision theorists call the $100,000 the "expected value" even though the only possible values are $1 million and nothing (so $100,000 is "expected" only in a specialized meaning of the word "expected"). As an additional complication, twice as much money is not always exactly twice as desirable, so it is more accurate to use something called utility, instead of money, in the calculations.

Mathematics aside, the ability to delay short-term gratification for longer-term rewards has been shown to result in higher achievement, even among young children. Teaching this skill would thus be a worthwhile goal.[9] Another approach is to advance the reward (i.e. the gratification) before completion of what it is supposed to reward. An example is paying a home improvement contractor for a new coat of paint up front, in part, and paying the rest upon completion of the job.

Upgrading "the system"

Even times far in the future will eventually become the present. Why not make those future times, as well as the times between now and then, as good as possible? In complicated terms, one could try maximizing the integral of the goodness of each time point from now to eternity, defining goodness of a time point as the expected utility of the probability density function of its predicted objective quality, with appropriate discounting of future utilities. But more simply, why not just ask your country and the world to be in as good a shape as possible in a generation or two instead of just next year? Ensuring that the world is in good shape when today's children come of age is not only intrinsically desirable, but also likely to be a good foundation for their children to inherit a good world, too. (It's an obvious conclusion, sometimes noted but mostly ignored.)

A world run for its long term benefit must be made of countries governed that way. Such countries must be based in turn on organizations that work that way, and those organizations must be made of people with the habit of long-term action themselves.

Constructing such organizations will require leaders and others who act long term.

But mainly, someone who successfully acts long term should be considered to have an important prerequisite for a leadership position and will have references and successors from previous jobs who verify it.

To summarize: Society should demand information about long-term results as a qualification for hiring or electing someone into a leadership position. The next few subsections give a few specific improvements.

"Collingridge's dilemma" and its practical application

David Collingridge identified an important difficulty, since called a dilemma, with regulating new technologies to best benefit society.[10] The dilemma is, when a technology is new, it is easier for government to control the technology's future impacts on society (e.g. with laws). But it is harder to foresee what those impacts will be and hence to act with the desired effects. On the other hand, as the technology becomes embedded in society, its negative impacts become easier to understand but harder to correct (because society develops financial interests and other internal forces that oppose corrective action). The dilemma, then, is that early on a technology is easier to control but harder to know how to control, while later when it becomes clearer what its problems are, it is harder to control.

Two examples among the many that fit the pattern are the use of personal data, and genetic engineering. Personal data has in recent years become broadly recorded and used. Video recordings are everywhere and pervasive companies like Google store and use a continuous stream of details about what people do and where they are. Privacy regulations might be desirable for protecting against government and corporate surveillance, but this may be a case of trying to shut the barn door after the horses have fled, putting this set of technologies on the "easier to understand, harder to control" side of

Collingridge's dilemma. On the other hand, genetic engineering is just getting started and will eventually probably rival or even exceed the centrality of computers in modern society. It is easier to set up a control regime now than it will likely ever be again, but what problems could arise that we could protect society from are hard to be sure of, because this set of technologies is early in its development and thus on the "easier to control but harder to know what to control" side of the dilemma.

Asking for a solution to Collingridge's dilemma is a bit paradoxical. I'd call it a "conundrum" except one of the reviewers of this book recommended trying harder to avoid unusual words. After all, it's not called a dilemma for nothing. So there is no magic bullet. As always, however, we can make the best of a tricky situation instead of settling for less. So here are a couple of ways to do that. (Note that "short term action" here means waiting until the problem impacts of a technology become obvious and pressing, and then attempting to control what is already a runaway freight train. On the other hand, "long term action" means looking ahead and anticipating problems ahead of time as best we can, and then putting down controls on the technology while it is still young, less entrenched, and easier to guide.)

- Countries can create special fast-track legislative and judicial processes for issues that are about new technologies. This would help avoid the multi-year delays that so often bedevil these processes, thus handling things early, before they become harder to control.
- People who are trained, and specialize, in foresight exist. This is a human resource that is not currently being employed in an organized way specifically to leverage their talents to anticipate future problems of new technologies.

Maybe you can think of some other way to better handle Collingridge's dilemma. If so, feel free to let us know.

The world culture that controls international relations

Let us shift our attention briefly from the overall good to competitive goodness: If your country's economy does better than economies of other countries over the next 20 years or so, then other countries will have less control over the well-being of your children. In the current and historical environment of amoral, cutthroat interaction among countries based on narrow national self-interest — a model of behavior that would earn a mental dysfunction diagnosis of antisocial personality disorder if displayed by an individual — that's worth thinking about. People with anti-social personality disorder tend to create chaos in the lives of other people around them. Maybe that helps explain the chaos of international affairs. The human race and a lot of individual people in it might be better off if the worldwide culture of cutthroat international relations was replaced by an alternative culture that expects more, if not perfect, synergistic cooperation, like what is customary in daily life among individuals.

Scenario simulation

To help people and organizations act long term, organizations can use scenario simulation, which involves investigating the implications of possible alternative futures. This approach is already much used in military planning. The natural next step is to legally require publicly owned corporations to create and maintain long-term strategic plans, as well as to use decision making processes that explicitly refer to them. In these plans, goals that run counter to the national interest could be disallowed, which is helpful because the remaining, desirable goals will tend to have a bit of extra influence by virtue of being written down. Additionally, the legal structure that governs corporate decision-making should be updated to encourage consideration of long-term factors. The value of being driven by long-term considerations is responsible in part for the success of Amazon, for example.[11]

Chapter Eleven

Battle for the Mind

(Chapter name quoted from the title of a book by W. Sargant, 1957)[1]

> *Can human memory be edited, revised, and changed? The evidence is in, and it can. Details range from frivolous to advertising applications to the collapse of the destructive "recovered memory" school of therapy. In the future, methods for revising human memory will improve. This should give us pause.*

In 1988, shocking accounts of memories of childhood sexual and even satanic ritual abuse began to sweep America.[2] Seemingly buried and forgotten for many years, these repressed memories were finally dredged up by adults under the encouragement, guidance, and leading questions of a controversial psychotherapy technique known as "recovered memory therapy." The book that started the movement was written by two authors, one a poet and the other a childhood incest victim, neither with training or background in psychotherapy, medicine, experimental psychology, memory research, or any other specific qualifications. Yet at its peak, recovered memory therapy had become a common practice among psychotherapists.

False memories

There was a major problem, however: Many of these recovered memories were false – despite the serious reality of childhood sexual and other abuse.[3] Far from being buried and then recovered, they were actually implanted by therapists, unhappy patients desperate for relief, and the expectations and over-active imaginations of both. Thousands of otherwise ordinary families were torn apart by false yet strongly believed accusations of incestuous and even mind-boggling satanic ritual abuse long ago. Many patients were made unhappier by this process rather than happier: Who wouldn't be put in turmoil by traumatic stories implanted in the mind and masquerading as real memories? The likely resulting alienation from parents and family might only compound the patients' difficulties.

Traumatic memories can indeed sometimes be repressed, forgotten, and later recalled. Yet false memories can also be implanted and made to seem real. The creation of a false memory is an important and unfortunate event: A "battle for the mind" has been fought... and lost.

Understanding memory implantation

That false memories can be implanted and then sincerely (but mistakenly) "remembered" may seem surprising. However it has been demonstrated in laboratory experiments. No one knows more about implanting memories than National Academy of Sciences member Elizabeth Loftus of the University of California, Irvine. Loftus's early research looked at the malleability of memories of vehicular collisions.[4] Working together, she and J. Palmer showed people a film clip of a vehicular collision. They then asked some people how fast the cars were going when they "smashed," and others when they merely "hit." One week later the experimental subjects were re-interviewed. Asked if broken glass was present, the results were dramatic. "Smashed" subjects were significantly more likely to remember seeing broken glass in the clip than "hit" subjects. Yet as Loftus and Palmer state, "There was no broken glass in the accident" shown in the film.

Numerous experiments deepened and extended the findings. In one test, a third of eye witnesses remembered an experimental "perpetrator" from a lineup that in fact did not include the perpetrator.[5] When falsely told that the perpetrator was in the lineup, the rate of incorrect remembrance climbed to over three fourths. One has to wonder how many innocent people are in prison because of this phenomenon. Battle for the mind: willingly lost.

In another experiment, Loftus & colleagues persuaded up to 18% of people that an implausible event — witnessing demonic possession as a child — probably happened to them personally.[6] Of course, that is different from falsely remembering such an event. Yet in a later experiment, 16% were persuaded that they remembered shaking hands with Bugs Bunny at Disneyland.[7] But the silly wabbit is a character owned by Warner Bros., not Disney! It definitely does not live in Disneyland. Harmless cartoon character fun? Perhaps. But things can get scarier. The obvious extensions to engineering false eye witness testimony in court, manipulating jurors, and political machinations are more than a little scary.

Overzealous and unprincipled prosecutors are bad enough, but history (and present practice, including in the US) shows that political partisans can say anything, do anything in attempts to manipulate public opinion. The Chinese government systematically attempted to alter memories of the 1989 Tiananmen Square massacre on a society-wide scale. Tools included doctoring photographs and inducing people to recount false versions of events that blamed the protesters. The power of this technique relies on the well-known capability of social pressure to modify memory.[8] Could memory manipulation happen in the US and other free countries? Vocal and wily political partisans will stop at nothing they can get away with, as any aware citizen knows. If they could alter society's memories to suit their purposes some of them would; if in the future they can, they will. You should care, since "to accept a false reality as truth ... is the very essence of madness."[9]

Hypnosis and hallucination

As a teenager I found an old book on hypnosis in the attic. Part overhype about its strange and awesome powers and part hands-on how-to, I was soon experimenting with a friend, who turned out to be a good hypnotic subject. Let's call him "Will." We lost touch decades ago but, concerning implanted memories, Will surely remembers this one. Under hypnosis I informed him that, upon awakening, there would be a mouse — a friendly one — which would hang around him, visible and unafraid. When he woke up from the hypnotic trance, sure enough, there was the mouse. Of course I explained that the mouse was not really there. It only seemed to exist owing to hypnotic suggestion, which he was fine with. Why not pick the mouse up, I asked later by phone. It will let you. Sure enough, it did. It will have a white stomach, I continued. Furthermore, its stomach will have an image of a rhinoceros on it, outlined in black hairs. Will picked up the mouse and reported that indeed it did have a white stomach with the black outline of a rhinoceros. Note that the stomach coloration and the rhinoceros were mentioned later, when he was wide awake and not under hypnosis at all. Regardless, those characteristics were there, implanted for Will to see and doubtless remember to this day.

In my case, a hypnotically induced hallucination occurred accidentally during high school class. My eyes glazed over as so often happened during lectures (and not just in high school). A hazy fog descended over my field of vision, probably caused at least in part by retinal fatigue. The teacher then uttered a sentence containing the phrase, "… write this …" of which I remember only those two words. My hypnotized mind took the phrase completely literally, and the word "this" slowly wrote itself in glowing, graceful cursive within the gray haze. I wakened in surprise as I realized this was a fascinating example of hypnotic suggestion. The hallucination disappeared immediately but the odd memory implanted by this hypnotically imagined event still remains.

Many people have memories not arising from objective reality. These memories can be vivid. Will's rhimouseros is one example. Such memories can arise from dreams and nightmares, as well as hallucinations. Hallucinations can be caused not only by hypnotic suggestion but also by psychoactive drugs like LSD, *Salvia divinorum*, etc.; psychiatric conditions like schizophrenia and severe mania; mystical and other religious experiences; and near death experiences (NDEs). Sometimes these memories are judged to be real, even uniquely transformative by the people who have them. Yet as well-known neurologist Oliver Sacks writes, "Hallucinations, whether revelatory or banal, are not of supernatural origin; they are part of the normal range of human consciousness and experience."[10]

It can happen to anyone

Leading memory expert Loftus herself was told by an uncle that, at the tender age of 14, she had tragically discovered her mother drowned in a swimming pool. The awful memory returned... but then the uncle admitted lying and "relatives confirmed that her aunt ... had found the body."[11] Battle for an expert's mind — lost. The flesh of the toughest soldier is no more resistant than yours, and minds can resist — or not — analogously.

False memories and the future

How effective will memory implantation get? As the future unfolds, will fake memories of vacations to Mars be mass marketed commercially, as in *We can remember it for you wholesale,*[12] the Philip K. Dick story that later became the motion picture thriller *Total Recall*? Will propagandists in politics and advertising revise our personal histories for us? Will psychotherapists do it for our own good, for example implanting memories of getting sick from ice cream to improve our nutrition and reduce obesity, as proved possible by Loftus?[13] Will it happen with our blessings? Will electronic chips containing "memories" of learning difficult subjects — foreign languages, history (revised or not), business and legal cases, technical, mathematical, and so on, be implanted in our skulls or "merely" piped into our brains via ultrasonic, transcranial magnetic, or pharmaceutical manipulation? Would you want them to be? What legal restrictions should be placed on memory manipulation? According to Loftus, "Over the next 50 years we will further master the ability to create false memories. ... The most potent recipes may involve pharmaceuticals that we are on the brink of discovering."[14] When we battle for our minds and memories, what counts as a win and what as a loss?

Recommendations

Cleverly and willfully used, advertising has been shown to be able to modify human memories. By its very nature, advertising will do this to serve the interests of the advertiser, not the interests of those whose memories are controlled. Consider the term 'propaganda': "A concerted set of messages aimed at influencing the opinions or behavior of large numbers of people."[15] Obviously those with power have always known about propaganda and how to use it. As Britain's Lord Acton (1834–1902) famously observed, "Power tends to corrupt." As obviously as it applies to politics, this definition also applies to commercial advertising, the typical purpose of which is to separate people from their money. When the almighty dollar rules, the ethics of memory revision... don't.

To help guard and protect memory, the "insidious mechanisms and methods" of memory modification should be taught in grade school, so that people learn to tell when it is being tried and thus can actively resist such mental manipulation.[16] Those who want to perpetrate memory modification will learn it regardless; the rest of us need to know about it for self-defense. For example, people should know that memories from adolescence are known to be the best targets for beer advertisers.[17] And so on. People should also know that memory revisionists "want the consumer to be involved enough [to] process the false information," yet not enough to "notice the discrepancy between the advertising information and their own experience."[18] Yechh. People should know that "imagining a childhood event inflates confidence that it occurred."[19] This imagining technique can be used to "recall" memories from infancy, before birth, and even as a sperm or egg cell![20] If that is not early enough, the same method suffices to recall "memories" from past lives.[21] As Loftus put it, "Memory, like liberty, is a fragile thing."[22]

Chapter Twelve

Will Artificial Intelligence Threaten Civilization?

You've heard the story: Intelligent robots try to take over the world. But this is only one possible future. One thing seems nearly certain: Artificial intelligence technology will continue to improve, and besting humanity in chess and Jeopardy! is only the beginning.

The first protest march against robots finally happened on March 14, 2015.[1] Although the Austin, Texas demonstration later turned out to be nothing more than a marketing stunt for a dating app, it garnered plenty of national media attention.[2] Clearly robots touch a nerve.

Humans are making computers, both robotic and stationary, with progressively greater intelligence ("AI"). Outplaying world chess champion Garry Kasparov way back in May 1997 was a convincing milestone. IBM's Deep Blue computer, the victor in that match, was the culmination of an effort that began with computers that lost, but ended with one that won. The general knowledge game *Jeopardy!* was another highly visible milestone achieved in 2011 when a computer playfully named Watson (again from IBM) beat the world's top players. Meanwhile, computer performance on a test of artificial intelligence, the classic "Turing Test," has been increasing over the years, with chatbot competitions held every year since 1991.[3]

If and when we finally make computers with intelligence superior to us humans, a critical tipping point will have been passed. The logic goes like this.

If human intelligence is capable of creating a machine of intelligence exceeding ours, then that greater intelligence should logically be able to design a machine of even greater intelligence. It could be a brand new machine or perhaps a modification to itself. This machine would in turn be able to design one with yet more intelligence. Continuing the process, a positive feedback loop would produce successively greater intelligence with no obvious limit. That is the essence of the AI singularity. Will it happen? The key question is whether humans are intelligent enough to create machines with greater-than-human intelligence. You could argue about what is meant by intelligence, but the concept itself is perfectly clear. If humans can't do it, then computers will always be useful — but no AI singularity. If we can, an AI singularity will occur, ushering in an era of rapidly escalating machine intelligence.

An AI singularity would be mysteriously unpredictable. Just as an animal species cannot know what the human era holds for it, we cannot know what such an era would hold for humankind. It would likely be weird... perhaps interesting... and whether for better or worse, a seismic shift in the human condition.

Let us assume that the current trajectory of increasing computer intelligence holds and that the singularity will in fact occur, and see what may happen.

The AI singularity is on its way

This would likely be within the lives of many readers because of the historically exponential growth trajectory in computer power. As noted first by Asimov in 1956 and later by Good, Vinge, Moravec, Kurzweil, and others,[4] when the AI singularity arrives current models of how AI affects society will become inapplicable, and what happens afterward we can only speculate about (like animals trying to think about what the human era means to them). In a practical sense that impossibility of understanding is what "singularity" means. Since we cannot know much about life under the singularity, what if something goes horribly wrong? Is that possible, or even likely? Yes, when one considers the tongue-in-cheek wisdom of "Murphy's Law."

Murphy's Law: If something can go wrong, it will.

This tongue-in-cheek heuristic, familiar to any engineer, is due to the innate complexity of most practical systems. This complexity leads to our inability to know how engineered systems will act prior to testing (or worse, using) them. Therefore, we need to be concerned about the dangers that might occur after the AI singularity. Because current models will no longer apply, we can't reliably assign probabilities, high or low, to the dangers. But we can be creative and try to identify all the risks we can — and then seek to protect ourselves from them.

Risks from AI arise from the mode of human-computer interaction (HCI) that occurs. We categorize the possible modes as cooperation, competition, and both combined.

- *The cooperation paradigm.* In this paradigm, AI will serve humanity as a new kind of tool, unique in part because of the literally super-human thinking powers it will have after the AI singularity occurs.
- *The competition paradigm.* According to this view — a sci-fi favorite — artificially intelligent entities will ultimately have their own agendas, which will conflict with ours.
- *Combined cooperation and competition.* In this possibility, artificially intelligent entities interact with humans cooperatively in some ways and competitively in others, leading to a confusing spectrum of possible futures.

These three paradigms each carry significant risks. The most catastrophic of these are outlined next.

Risks from the cooperation paradigm

These risks can be insidious. They involve "killing with kindness." Who wouldn't be tempted to rely on a legion of intelligent robots with awesome powers to make life a paradise? The possibilities are endless. Early AI pioneer John McCarthy of Stanford University quaintly proposed a robotic closet that "… a woman may step into … which will … put on clothes most suited to the occasion and fix her hair …"[5]. (The complete quote is too odd to reprint here. It also fails to explain what happens if a man, by accident or on purpose, steps into such a closet. For example a burglar trying to hide. Or what about a kid in a game of hide and seek? Or a pet dog or chimpanzee…) Be careful what you wish for.

Robots imbued with artificial intelligence could potentially eliminate the emotional need of individuals to interact with other people, leading to social and perhaps population collapse. As early as 2003, NEC was selling PaPeRo, a nannybot for kids — shades of the classic sci-fi stories by Isaac Asimov ("Robbie") and Philip K. Dick ("Nanny").[6] Laws against robots being made with certain key human-like characteristics might be a sufficient safeguard. What such laws would be effective? We need to find out before it is too late.

Sufficiently capable robots could build more of themselves. Artificial intelligences that are made only of software could of course be duplicated as easily as any software. Myriads of alien intelligent entities with "intellects vast and cool and unsympathetic"[7] could come into existence, until as many existed as people wanted to do their bidding, if the "invisible hand"[8] of economic forces has its way. Those forces might lead these bots to efficiently farm, mine, and do other activities that affect the natural environment. Such armies of bots could damage the environment and extract non-renewable resources considerably more efficiently than humans. The economic paradigm that the world currently operates on incentivizes this tendency, making it potentially difficult to prevent.

To solve the problem of robot hordes wreaking havoc at our own bidding, other economic paradigms are needed that incentivize stewardship of the Earth, rather than its exploitation. What might such alternative economic systems look like? Much more remains to be discovered about this complex question. Since humans are already damaging the Earth without intelligent robots to help, creating new economic systems might preserve the ability of the Earth to provide comfortably for the needs of billions of humans, after the AI singularity (if and when it happens) as well as before.

That brings us to the next major risk. Such robots could make it unnecessary for humans to do or be very much. The analogy in the world of biology is parasitism, and the risk for us is underlined by that fact that, probably, most animal species are parasites.[9] Coddled by automation, our species might suffer insidious cultural and genetic deterioration, because "... parasites often become extremely simplified ..."[10]. What forms could such deterioration take, how far it could go, and how would we recognize it when it occurs? What are the solutions? These are challenging questions that some readers may wish to try to solve.

Risks from the competition paradigm

These risks are a perennial favorite of apocalypse-minded science fiction. The robots become autonomous and make their move. Humans run for cover. The war is on, and it's them or us, winner take all. If bots do take over, any remaining humans risk having little choice but to hide, while cold metal robotic machines methodically "roboform" the Earth to make it suitable for bots but unable to support human life. For example, oxygen causes rust and this is bad for robots, so they might try to get rid of it. Guarding against such risks is a tough proposition: We don't know how to do it. But we do know how to think and discuss. So we should do that, hoping that such a foray into a forward-thinking existential robology will be able to find solutions before it is too late.

Risks from combined cooperation and competition

Artificial intelligence could be embedded in robots used as soldiers for ill. Such killerbots could be ordered to be utterly ruthless. Just as weapons of mass destruction like nuclear bombs and biological weapons are threats to all humanity, robotic soldiers could end up destroying their creators as well. The movement to ban AI-enabled weaponry finally began in earnest with an open letter to the public on July 28, 2015.[11] It was signed by thousands of AI and robotics specialists as well as luminaries like Stephen Hawking and others. I predict that the movement to ban weapons that choose their own targets will pick up support and become like the movements to control other weapons of mass destruction: chemical, nuclear, and biological weapons.

Another example combining cooperation and competition is a highly capable AI tasked with promoting the interests of a large organization, such as a corporation or other special interest group, even to the detriment of the rest of society.

An emergent property of a corporbot, robosoldier, or other AI intent on doing what it was created to do as effectively as possible would, logically, be a drive to improve itself. By seeking to increase its own intelligence and other capabilities, it would be more effective, exactly what it was built to be. It would then become increasingly able to pursue its assigned goals by unanticipated — and perhaps highly undesirable — means. For example, there exists no surer, more permanent way to end the common cold than vaporizing the biosphere, including all human inhabitants, many of whom would have expected long and cold-free lives. The risk of unanticipated and potentially disastrous side effects highlights the urgency of the classic advice "be careful what you wish for." This was explored in classic tales from "The Monkey's Paw" and the venerable bottled genie bearing wishes,[12] to numerous robotical yarns from Shelley's

Frankenstein[13] to Asimov and his three laws of robotics[14] as well as many others.

What else can we do?

While some suggestions are provided above, clear solutions are hard to come by and, so far, have not been agreed upon. Simply letting over-enthusiastic technology developers go their merry way without some caution tends to lead to fears of existential risks for humankind. Unfortunately the question of how to guard against destroying ourselves with such technologies does not seem to be resolving. Therefore it is time to seriously address the meta-question of why. Why, indeed? Answering that question might ameliorate existential risks, a very practical goal, as well as shed light on foundations of the human condition.

Chapter Thirteen

Deconstructing Nuclear Nonproliferation

Nuclear weapons are spreading to more countries over time. Where can we expect this trend to lead?

Little-known Soviet military veteran Stanislav Petrov was finally honored in May, 2004 for single-handedly heading off nuclear war between the United States and the Soviet Union. Alerted on an ordinary September day in 1983 that a US intercontinental ballistic missile was heading for the USSR, he guessed correctly that it was a false alarm, and declined to launch a nuclear counterattack.[1] Hardly a stable situation. If "deconstructing" means identifying internal contradictions, nuclear proliferation is a good example. Let's examine this a little more closely.[2]

The first country to go nuclear was the US, with a test explosion in 1945. The Soviet Union tested their first device in 1949, followed by the UK (1952), France (1960), China (1964), India (1974), Israel (probably 1979), South Africa (probably 1979), Pakistan (1998), and North Korea (2006). Regardless of the uncertainty of a 1979 test, South Africa definitely did develop nuclear weapons, which they later destroyed, and few dispute that Israel has them although the government will not confirm it. Thus we see nuclear weapon development programs reaching "fruition" (the seeds of destruction yielding dangerous fruit) 10 times in 70 years by 2015, averaging once every 7 years and with intervals ranging from 0 to 19 years. It turns out that 8 out of 9 of these intervals are 10 years or less, suggesting statistically that the next country to build a working nuclear device is likely to do so within the next several years.

Iran is often feared to want nuclear weapons. This was a key concern in the negotiations that resulted in a nuclear agreement with

Iran on July 14, 2015. As a factual matter, however, a country cannot want; rather, individual people can want. Common sense suggests that some individuals in Iran want them and others do not. The important question is thus what the people in charge there want. A further uncertainty is technical in nature and due to the ambiguous purpose of gas centrifuge enrichment, which extracts uranium isotope 235 from its less useful 238 isotope. U-235 is useful for both civilian applications and nuclear weapons, depending on its level of purity. The military potential of U-235 enrichment led Iran into conflict with the US, Israel, and many others.

When it comes to advancing along an enrichment path that could lead to nuclear weapons development, neither Iran nor its opponents played games. Iran has treated its people suspected of causing problems with considerably less kindness and consideration than the west has treated its own nuclear weapons opponents, such as J. Robert Oppenheimer, Julius and Ethel Rosenberg, and Mordechai Vanunu. Executions and disappearances have been part of this. For its part, the West has also acted harshly, including apparent assassinations of Iranian nuclear engineers. Becoming an engineer in Iran's nuclear program has been a rather risky career move.

Perhaps the community of nations can succeed with Iran as it has with other countries that have backed off enrichment programs, such as Argentina and Brazil, or even have technically owned weapons, like Ukraine, Kazakhstan, and Belarus. The ultimate picture, years downstream, will become clearer in due course.

Mathematics of the future

"It is difficult to predict, especially about the future"[3]; past performance does not guarantee future results. However if past trends in nuclearization continue then the number of nuclear nations will approximately double within the lifetimes of today's children. If unchecked, that number would continue rising. Such a rise is serious in part because a nuclear country could potentially attack a non-nuclear one, a crisis both for the victim as well as for international custom and culture. Will the international response be mostly hand-wringing, which would effectively be an incentive for future such attacks, or would it make the aggressor nation wish it had not attacked? Hand-wringing would encourage non-nuclear nations to nuke up in self defense, making the world still more dangerous. However, making an aggressor sorry it had attacked can be a nasty and dangerous business.

Mathematically, doubling the size of the relatively small nuclear club nearly doubles the number of possible such attacks on non-nuclear parties. For example, 10 nuclear countries and 200 non-nuclear ones pose 2,000 possible attacks since each of the 10 could attack each of the 200. Twenty nuclear countries and 190 non-nuclear present 20×190 or 3,800 attack pairs, almost though not quite double.

This unfortunate trend is self-limiting, because when the number of non-nuclear nations starts to shrink appreciably because they have transitioned to nuclear status, the number of one-sided attack possibilities begins to level off. If enough countries transition to nuclear status then the number of possibilities begins decreasing. Ultimately if all nations go nuclear the possibility of a nuclear attack on a non-nuclear nation declines, of course, to zero. Yet this hardly justifies a sigh of relief because at the same time the number of possible nuclear exchanges between two nuclear countries increases. In fact it not only increases, it accelerates, increasing faster and faster with every new member of the nuclear club. Here is why.

Given 10 nuclear countries, adding an 11th means adding 10 new possibilities for a nuclear exchange, those being between the new 11th and each of the original 10. Add a 12th, and the new possibilities this creates is more, not 10 but 11, one between the new 12th and each of the preexisting 11. You can actually count up all of the possibilities by starting from the first two nuclear countries (which gives one possibility of a nuclear exchange) and progressively adding a third (which adds two possibilities, between the new country and the pre-existing two), a fourth (which adds another three possibilities), a fifth (adding four), etc. Technically speaking, the total number of possibilities is pretty close to the square of the size of the nuclear club (that is, the size × the size), halved. Thus, while 10 nuclear countries have about 10×10/2=50 possible lines of attack, double that to 20 countries and now there are about 20×20/2=200 attack possibilities: Doubling the club means 2×2 or 4 times as many potential nuclear exchanges, tripling it means 3×3 or 9 times as many, and so on. One hundred nuclear countries...about 100×100/2=5,000 possible attacks. That's 10× as many countries but 100× as many possible nuclear exchanges; and the world has more than 100 countries.

Not convinced? Then simply wait: If the probability of an attack over the next year is low, the probability over the next two years is approximately doubled. Pick any probability you like for a typical year, wait enough years, and although the size of the increment in probability added per year shrinks, the total probability gets closer and closer to 1 — that is, certainty.

What can be done

If knowledge is power, then better understanding of proliferation will lead to more power to control it. To better predict what the statistics of the past say about the future of nuclear proliferation, we need to better understand, statistically, what the historical trends really are. We already know the next country to go nuclear may well have a device within several years. This is a rough initial observation; more detailed analyses would produce more detailed prediction hypotheses.

What's needed is a metric for quantifying the rate of nuclear proliferation. Limiting such a metric to counting full members of the nuclear club is coarse-grained, like a tape measure marked off in feet but not inches. A finer metric would also count steps toward nuclearization as appropriate fractions of full membership. These include acquisition of related prerequisites like centrifugal isotope purification equipment, fissionable materials like uranium, and device delivery systems. Members of the nuclear club that proceed further to development of fusion-boosted devices ("H-bombs") and more complex two-stage devices should count extra, as should other second generation devices like neutron bombs. Other factors to include are developments and installations of monitoring technologies, and nuclear weapon deployments in non-owner host countries. Such hosts are of more than academic interest. For example Cuba's status as such a country brought the world close to a second nuclear war in 1962 (hence the term "Cuban missile crisis").

Disarmament events could also be worked into a metric. For example, South Africa developed its own nuclear weapons but then was the first to voluntarily disarm. A second set of cases followed the breakup of the Soviet Union. Ukraine exemplifies an interesting example. That nation was briefly the owner of the third largest nuclear arsenal in the world, having inherited a large quantity of operable nuclear devices from the Soviet Union upon its breakup. Fortunately it was willing to return the weaponry to its original *de facto* owner, Russia. A similar situation occurred with Kazakhstan. While that involved fewer (though still a lot of) weapons, there were more international fears about their disposition and security. Belarus constitutes a third post-Soviet case, and was for a time reluctant to return the weapons, though was finally persuaded to do so.[4]

Breaking the trajectory

A metric based on integrating the kinds of ambiguous and partial nuclear weapon ownership data points outlined above would go a long way toward realizing the goal of accurately measuring the world's rate of nuclearization. Additionally, social and political factors associated with nuclearization should be formalized for use in a metric to the degree this is possible. These are particularly important because they can also provide guidance in avoiding nuclearization; technical and statistical considerations can only go so far in understanding and controlling the problem.

If we get at and modify the causes, we can break the statistical trajectory, preventing or reducing proliferation. A useful place to start is by identifying the causal "pressure points" of the system. Key such causes include the incentives of being able to (a) attack, (b) retaliate to an attack, (c) threaten to attack, (d) reduce the threat of attack, (e) sell nukes, (f) gain prestige, and (g) etc. If we change the incentive structure appropriately, the improvements could mitigate the causes of proliferation. Naturally this is not a trivial task. How, then, to do it?

Former US Secretary of Defense Caspar Weinberger was once asked in a televised news conference by a concerned listener why the reasons for nuclear proliferation were not investigated more deeply. His response: The ideas in new writings were simply repeating those in existing ones, and what is the point of more research when it would just be redundant?[5] In other words, he advocated that we give up rather than direct more research to questions like, for example, *why* adequate understanding was so elusive. The shortsightedness of his reply was surprising in one chosen to be responsible for the existential safety of his country. Suppressing the search for an understanding so important to the existence of one's country, civilization, and perhaps even the entire human race makes such attitudes not only illogical and incompetent, but downright dangerous.

The Second Generation:

The Next Thousand Years

Chapter Fourteen

Space Empire — From Mercury to Neptune and Beyond

The lure of interplanetary colonization is intriguing. Let us examine the possibilities for various planets and other bodies. Mars is interesting, but it's only one possibility.

It is a refreshing fact that the prospects for human survival are substantially higher if we live on two worlds, instead of just Earth. The Moon, say, or Mars... every extraterrestrial body poses unique technical challenges to colonization. Yet nearly all are potentially habitable — in theory. Our survival prospects climb higher for three worlds and higher still for four, because if three fail the fourth lives on, whether it is on Earth, the moon, or anywhere else that we can get to. It's like flipping quarters: The more you flip, the greater the chance one or more will come up heads. We will start near the Sun and work our way out.

To stay grounded, bear in mind the enormity of the technical challenges inherent in off-Earth colonies. Maintaining sealed, self-sufficient human environments has been tried on Earth, notably in the Biosphere 2 project near Tucson, USA, and the BIOS 3 project in Krasnoyarsk, Russia. But the technology still remains far from perfected. If the problem is not yet solved even here on Earth, how much more difficult it must be to make such an environment work extraterrestrially! Still, the presence of water is helpful and, as NASA official Jim Green put it in 2015, "The solar system is ... a pretty soggy place."[1] Thus while the following accounts are not going to happen in a mere year or two, longer term they are certainly plausible and worth working toward.

Colonizing Mercury: Major roadblocks and their solutions

Mercury is typically underestimated as a possibility for colonization, though it certainly would not be easy. Let us begin by giving our putative future settlement a name — Mercuria, because it is on the planet Mercury. Mercury is the closest planet to the Sun and has no significant atmosphere. So to found Mercuria, we must solve the basic problems of heat and vacuum. These problems apply both to the trip there and to the colony itself. Space ships handle the vacuum problem nicely by providing an airtight enclosure to live in. Appropriate shielding, such as a thin reflective shroud, can protect against surprisingly intense heat sources in space. Given such a ship it seems natural to use it, after landing it on the planetary surface, as the initial structure housing the colony itself. Later, domes with a view could potentially be built. However digging underground caves, although challenging even here on Earth, seems easier as it would not require special dome-building materials that could hold air pressure with the necessary near-100% reliability.

Some places on the planet provide better prospects than others for placing our new community of Mercuria. Water exists in large quantities on Mercury in the form of ice, typically hiding in a layer just a few inches underground in some areas.[2] Mercuria should naturally be built at a location near one of the ice deposits. You might wonder how a planetary surface that regularly reaches around 800 degrees Fahrenheit could contain ice — well, some craters near the poles contain areas in perpetual shadow. This is because sunlight hits the pole areas at very oblique angles, shading many low points in the craters. Without an atmosphere to provide a hot wind from locations at 800°F (425°C) into the shaded craters, these craters stay extremely cold.

Although Mercuria should be put near plentiful ice deposits, it wouldn't do to have Mercurians shiver or even freeze from the cold, especially being on such a hot planet so near the Sun. Warm clothing will only go so far. However, the intense sunlight outside the shadows provides abundant solar energy that could be tapped as a source of heat and light. This would provide the energy needed to warm the community of Mercuria as well as to support growing crops, raising edible algae in tanks, or generating electricity to support manufacture of human nutrients using chemical engineering processes (yum?).

An ideal place to collect solar energy would stay lit 24/7 (or the Mercurian equivalent, keeping in mind that a day-night cycle on Mercury lasts two Mercurian years, or 176 Earth days).[3] The poles, in addition to being cooler than other places, have a few such special spots, mountain peaks that receive sunlight all or almost all of the time. This is important because nighttime on Mercury is as long as daytime — another 88 Earth days. One wouldn't want to be without power for that long. These mountain tops are called *peaks of eternal light*. The perfect spot for Mercuria would thus be in an icy "pit of eternal shadow" at the bottom of a crater, near a "peak of eternal light" at the top of a neighboring mountain.

Once the new community of Mercuria gets underway, population should begin a steady increase. Offshoot communities will soon be needed at other hospitable locations. I say soon because at a population increase rate of 2% per (Earth) year, 100 Mercurians will become a million in just 467 years — and ten billion in another 467 years. Successively less ideal locations will need to be colonized, using successively more well-developed methods for taming progressively harsher conditions. Assuming that unlimited cheap solar energy is sufficiently enabling, Mercury is actually a great place to be.

A scenario for Mercuria

Let's envision the growth of Mercuria. Colonists would dig underground caves, perhaps using equipment similar to that used for horizontal oil well drilling on Earth. Caves keep a steady temperature and can hold a breathable interior atmosphere much more securely than a fragile bubble. There they will experience cramped, crowded conditions because of the difficulty of digging as the automated drilling equipment brought to Mercury, if any, wears out. But all colonists would spend substantial amounts of time with a pick, breaking off pieces of underground rock and soil to enlarge the cave. Sunlight from a nearby peak of eternal light would be used to illuminate the gradually expanding cavern in order to grow crops. This light could be obtained by using numerous mirrors to reflect the intense sunlight onto a translucent portal at the entrance to the cavern. That spot would then shine brightly on the underground cavern like its own private little sun.

The local manufacture and use of these mirrors would be a technical challenge for which Mercuria's schools would prepare its youngsters with an intense science and engineering curriculum. Mirrors are an appropriate strategy because they are relatively easy to make. Their production does not require the more extensive industrial infrastructure needed here on Earth to sustain other lighting strategies. However, swiveling of the mirrors would be needed, to keep the sunlight directed onto the entrance portal as the planet turns. Fortunately that would be eased by the very slow motion of the Sun in the sky (one day-night cycle being 176 Earth days long). A low tech but locally sustainable strategy would use long, hand-cranked rope-and-pulley assemblies. Kids would slowly operate these mirrors from within the cavern while working tricky homework problems on the math, chemistry, physics and biology of ropes, pulleys, mirrors, planetary motion, and plants. Sure, computerized controllers could do it too. The problem is that manufacturing computer chips is such a challenge even on Earth that it seems unlikely on Mercury during the pioneer stages of colonization. It'll be hard enough just to grow food plants. Kids would need to include plant studies in their studies, to support the colony's need to keep the interior of the cavern densely grown with carefully tended food plants.

Much of the creative energy of the colonists would be devoted to developing indigenously sustainable ways to electrify and automate the mirror swiveling task, an effort that would be increasingly successful over a span of a few generations. Careful choice of Mercuria's location coordinated with its energy usage would keep the cavern temperature steady and comfortable, rendering individual dwellings within the cavern unnecessary. People could simply stake out a spot near a wall for a small, flimsy lean-to in which to sleep and store their few personal possessions. They could call it a "nest," and children would likely look forward eagerly to the day when they are permitted to set up their own. Life in the tiny, cramped village would have an intense and small-town cultural flavor very different from what most readers are used to.

After a few generations of population increase, the cavern would reach a maximum size. Cavern size is limited by the ability of the sun-like portal to light and sustain it. Therefore colonists would have to build new, connected, yet distinct caverns with their own surface portals and mirror systems. Disconnected communities would also need to be established to dig habitation caverns under different suitable craters. With a likely 2 or 3% annual rate of population increase, it would take only a few hundred years for Mercury to host a thriving civilization. That in turn could support much higher technical capabilities than the early Mercuria. Therefore as colonization technologies improve over the generations, a diverse manufacturing infrastructure would develop amidst communities with complex specializations and trade relationships. The distinctive, rigorous early pioneer lifestyle would gradually become more cosmopolitan and even comfortable.

As ideal cavern locations become harder to find, caves at locations with warm or chilly average temperatures could be made habitable by cooling them with air conditioners or warming them with heaters. The habitable areas of Mercury will soon become over-populated, posing a problem that needs to be solved, just like on Earth.

Colonizing Venus: Whoa!

There is a surprising undercurrent of interest in colonizing Venus, even though a less likely place could hardly be imagined. As of this writing, Google notes several times more hits on the query "colonizing Venus" than on "colonizing Mercury" (even though Mercury is far more hospitable, bleak though it is). Even the query "colonizing Earth" gets only about half as many hits as "colonizing Venus," though the number of people actually colonizing Earth (7 billion), greatly exceeds the number colonizing Venus (zero). Venus is about twice as far from the Sun as Mercury, therefore receiving only about ¼, or $1/(2 \times 2)$ as much sunlight intensity (the so-called "inverse square" law). Despite this, Venus is actually hotter than Mercury because it suffers from a major greenhouse effect.

Like on Earth, the greenhouse effect on Venus is due mostly to carbon dioxide in the atmosphere. But on Venus the carbon dioxide concentration in the atmosphere is much greater, 96.5%, which is about 2,400× greater than Earth's mere 0.04% or so. Furthermore Venus's atmosphere is much denser, with a surface pressure about 92× of that on Earth, so the actual density of carbon dioxide at the Venusian surface is 92×2,400 or over 220,000× greater than Earth's. This explains why Venus has a much more severe greenhouse effect than Earth does. Indeed, Venus's greenhouse effect makes its surface fairly steady at about 860°F (460°C), day or night, equator or poles.

That's hot — quite a bit hotter than a self-cleaning oven, in fact. Fortunately the extreme runaway end point of the greenhouse effect on Venus is not a concern here on Earth.

Colonizing Venus would be much harder than colonizing Mercury, the moon, or Mars. The surface is so hot that at night it would glow faintly. During the day the glow would be overwhelmed by what sunlight filters through the dense cloud cover (Figure 3) to reveal a rather desolate landscape. Protecting colonists from the heat would require heavily insulated dwelling spaces as well as heavy-duty (and *very* reliable) air conditioning. The high atmospheric pressure would require either very strong walls or a special gas mixture suitable for humans at that pressure, perhaps argon based with a suitable amount of oxygen mixed in. The most reliable and doable approach might be to build underground.

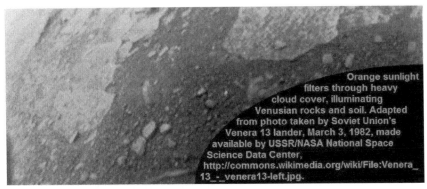

Orange sunlight filters through heavy cloud cover, illuminating Venusian rocks and soil. Adapted from photo taken by Soviet Union's Venera 13 lander, March 3, 1982, made available by USSR/NASA National Space Science Data Center, http://commons.wikimedia.org/wiki/File:Venera_13_-_venera13-left.jpg.

Figure 3. Venusian surface in daylight. Yes, image is tilted.

To get the energy to run the air conditioning and so on, realistic (and I use the term loosely given current engineering capabilities) sources could include the following: high temperature solar panels able to work under the dim, overcast light conditions of Venus's daylight period, which is about 58 Earth days long; windmills, which should work well in Venus's thick atmosphere; and an energy storage or other emergency backup system that would activate during

Venus's nighttime period, also about 58 Earth days long, if the surface should be becalmed thus inactivating the windmills.

As for leisurely strolls on the surface, that would be a real technical challenge but would increase the attraction of living on Venus. Speaking of attraction, suitable rites based on the Roman goddess of love Venus, after whom the planet is named, could also make life there more interesting.

Going airborne

Fans of Venusian colonization — from whose ranks would presumably come the necessary volunteers — could find temperatures and pressures tolerable for us humans roughly 55 km (35 miles) above the surface. Hence, the city-in-a-humongous-balloon concept. A 2-mile (3 km) diameter balloon containing a breathable atmosphere, with oxygen generated by plants or solar energy from the ambient carbon dioxide, would generate millions of tons of lift in the denser, carbon dioxide-rich Venusian atmosphere at habitable altitude.[4] That's enough lift to carry a reasonably sized self-sustaining colony inside the balloon. Best of luck! Just don't fall out, because after diving through 55 km of dense, superheated gases there might not be much left by the time your smoking remnants hit the equally superheated, but even denser, ground.

The Moon

From the Earth to the Moon (1865) is a novel by science fiction pioneer Jules Verne (1828–1905). It is mentioned in H. G. Wells' later novel *The First Men in the Moon* (1901). The explorers in Wells' tale found huge "mooncalves," breathable air and extensive underground cities, but no green cheese. Modern colonizers will instead have to contend with a more Mercury-like environment, though the Moon is closer, cooler, and has less interminable days and nights. It is also easier to get to. This has been a given since July 20,

1969 at 9:56 p.m. Central Daylight Time, when astronaut Neil Armstrong stepped onto the lunar surface and proclaimed "That's one small step for [a] man, one giant leap for mankind." (He left out the "a," as is clear from the audio recording at www.nasa.gov/wav/62284main_onesmall2.wav.)

Let us call the first moontown "Luna," and its citizens "Lunarians" ("loonies" would be considered off limits in polite company). How would the problems facing the Lunarians necessarily differ from those of the Mercurians?

Daytime on the Moon lasts an average 13.66 Earth days, several hours shy of 2 weeks. Same for nighttime. Since on Earth a fortnight is two weeks, Lunarians can just dispense with the term "night" and use "fortnight" instead. This is a quite different experience from here on Earth and different from Mercury as well, where nighttime drags on for 88 Earth days. One approach to finding a place to hang your hat on the Moon is mooncaves (not mooncalves, which would probably eat any such hat if they existed).[5] Astronomers believe there are enough caves, called lava tubes and created by ancient flowing lava, to give colonists choices. Although living in caves might not sound super-advanced and high tech, keep in mind these caves are on the Moon (and as the saying goes, "be it ever so humble, there's no place like home" — even if it is on the Moon). Alternatively, for an artificially warmed and above-ground colony with access to ice for water, the same strategy for an ideal location applies as for Mercury: a pit of eternal shadow inside of a polar crater adjacent to a peak of eternal light. The best currently known location for Lunaria is the Mount Yewridge area near the South Pole.[6] A spot on Mount Yewridge is almost permanently lit, and the mountain is within the polar area, situated in a large depression containing plenty of areas of eternal shadow.

Finally, the Moon is a lot closer and easier to get to than Mercury. Therefore it is correspondingly easier to build on. Thus the

Moon colony of Lunaria can expect to be founded much sooner than Mercuria, perhaps by your children or even yourself.

Why not colonize Earth?

The obvious objection is that we've already done it. Yet, consider the vast uninhabited areas ripe for colonization. Huge deserts. Frigid Antarctic wilderness. Vast ocean surfaces! Sealed domes in the deserts are the easiest. Floating sealed domes are next in difficulty, while Antarctica is perhaps the toughest challenge. If self-sufficiency breaks down, colonists could return to civilization with relative ease, compared to failure of an off-planet colony which would mean almost certain death. A further advantage of developing self-sufficient colonies on Earth is the opportunity to get a start on some of the same technologies that would be needed in off-planet colonies.

The Martians

A long-standing cultural fascination with Mars helps make it an attractive colonization option. Famed sci-fi author Ray Bradbury, in his book *The Martian Chronicles* (1950), described children on Mars curiously asking about Martians, then looking into a canal and seeing "The Martians … reflected in the water." The first Martian colony may be named "Chroniclia" after that book (alternatives like "Mars Colony 1" will likely be considered too boring). Its first leader? Let us call him "Chroniculus I the Fearless" (more stirring than his birth name, which might be Bart Smith).

Chroniclia will no doubt be established near the Martian equator. The poles are too cold. Even if you wouldn't mind seeing dry ice "snow" through the transparent dome wall, it would be hard to keep the dome comfortably warm, especially at night, even with a double-walled dome designed to retain warmth. Better to put such a dome at the equator, where the surface temperature can get as high as about

70°F (18–21°C). A location near frozen underground water will be helpful. Even liquid water has been found on Mars![7] At night the surface goes below −100°F (−73°C), so the dome will need to be able to trap and retain solar heat overnight, perhaps using water pools or the ground for heat storage. The dome also needs to be airtight and strong enough to hold pressure, because the Martian atmosphere is extremely thin, with an average surface pressure less than 1% (more precisely, 0.6%) that of Earth at sea level. By comparison, pressure at the top of Mount Everest is about 33% sea level, barely enough to sustain human life for very limited time.

On a positive note, the atmosphere is 95% carbon dioxide, which can be pumped into the dome as needed to support crop growth (plants "breathe" it) and manufacture of carbon-based materials like plastics. The most noxious component in the atmosphere is carbon monoxide, at 700 parts per million. Carbon monoxide is dangerous at that concentration, so pumping in much Martian air would necessitate measures to remove the carbon monoxide.

Let us also give our future Mars colony a legendary second leader, who we name Chroniculus II, known to succeeding generations as Mother of Mars (MOM for short). Under MOM's guidance Chroniclia may perfect dome construction of tough, transparent, locally manufactured plastics much easier to make than glass. Population increase will motivate using such technology to build new domes — and fast — as a plausible steady rate of population growth (such as 2%/year) will crowd the Martian surface with a teeming civilization in a matter of under a thousand Earth years.

A few other characteristics of Mars will lend local color to life there. One is the knowledge of Earth's long cultural fascination with Mars.[8] That will make colonists feel important. A more practical consideration is that the Martian day is 24 hours, 39 minutes, and 35 seconds long. This is perfect — much better than Earth for the many of us who would appreciate waking up about 40 minutes later every morning than we did the previous morning.

On the other hand a Martian year is roughly double an Earth year, so holidays will be a lot less frequent unless more of them are instituted. Maybe the birthdays of some planetary heroes could be celebrated (Chroniculus I day, MOM's day, etc.). Also the gravity is only about 37.5% of Earth's, so if you weigh 165 lb. (75 kg) here on Earth, on Mars you would weigh 62 lb. (28 kg) and could jump around like a hyperactive kangaroo. That would be fun. But the potential downside of such low gravity is its long-term effects on health, which are not known. Required, rigorous daily calisthenics might mitigate some of those problems, but that might reduce the number of people interested in becoming colonists.

Ceres: Half a planet is better than none

You might not have heard of Ceres (pronounced like "series"), but it exists, in the asteroid belt between Mars and Jupiter, and as a colonization option it merits serious consideration. The largest asteroid, it is officially classified as a "dwarf planet," — yet may have more fresh water (as ice) than Earth. With a diameter of 590 miles (950 km) it is indeed dwarfed by Earth's 7,920 miles (12,750 km), thirteen times as much. Surface area is more relevant to colonization than diameter, however, and just as a foot has 12 inches but a square foot has not 12, but 12×12 or 144 square inches (similarly 1 m² has $100^2=10,000$ cm²) the surface area of Earth is 13×13 or 169 times greater than that of Ceres. Yet, if you stood on its surface it certainly wouldn't feel small. Geometry tells us that to a person standing on the surface, Earth's horizon is only a little over twice as far as Ceres'.[9] That's enough to give a similar sensation of distance, even though this dwarf planet is so much tinier than Earth.

On the other hand, other facts of day-to-day life on Ceres will be downright out of this world compared to our experiences on mothership Earth. Ceres rotates every 9 hours, 4 minutes, and 27 seconds. Since a Cerean day is just a few minutes over 9 Earth hours, for convenience the Cereans may define their hour so there are exactly 9 of them in a Cerean day. The tiny difference between an Earth hour and a Cerean hour (roughly half a minute) would be imperceptible.

Life with days 9 hours long would be an interesting experience. A circadian sleep-wake cycle within that time frame is probably ridiculous, but a 3-day cycle of 27 hours (Cereans will call it a "triday") might work nicely. You would rarely have trouble getting up, since the 27 hours compared to Earth's 24 means that getting up late by Earth's standards would still be early on Ceres, so you could get a few extra things done before leaving for work. With everyone living in bubbles or underground, the commute would be shorter too. A typical Cerean schedule (Table 3) should help you prepare to move there (courtesy Ceres Bureau of Tourism and Immigration).

Very low gravity

With gravity only about 1/36 Earth's, things are different indeed. Try this now: Jump just hard enough to get a mere 1 inch (2 ½ cm) off the ground. On Ceres, that would get you about 3 feet (1 meter) up. Can you jump up a foot? On Ceres you'd fly up 36 feet (11 m) unless you hit the ceiling first. Walking around, one would tend to bounce off the ground with each step, which might resemble a flying leap more than a step. Running would be effortless and fast. It would probably feel a lot like flying. For the grueling around-the-asteroid trek that groups of young Cereans (called "trekkies") might someday take before being welcomed into adult society, they would be trained to swing a long trekking pole at the ground, the tip hitting obliquely at about 20 miles (30 km) per hour. This would propel the youths forward at that speed,[10] and up enough so their feet never touch the ground.

115

Table 3. Typical Cerean resident's schedule

6:00 a.m.: Morning sunrise. Jump out of bed (gravity is low, and you are well rested).

9:00 a.m.: Arrive at work.

9:59 a.m.: The hours only go up to 9, unlike the 12 on Earth, so in another minute it will become 1:00.

1:00 m.d.: Ante-midday (a.m.) period transitions to midday (m.d.) period.

1:30 m.d.: Midday sunset.

3:00 m.d.: Firstlunch.

3:30 m.d.: Back to work.

6:00 m.d.: Midday sunrise.

6:30 m.d.: Secondlunch.

7:00 m.d.: Back to work.

9:59 m.d.: Work will be over in another minute.

1:00 p.m.: Work over. Transition to post-midday (p.m.) period. Go home.

1:30 p.m.: Evening sunset.

5:00 p.m.: Bed time. You've been up 17 hours, and it is easier to fall asleep while it is still dark. Your Earth-evolved 24-hour clock is perpetually trying to catch up to the Ceres 27-hour triday cycle, making you sleepy now, yet alert by the time to rise. You are ready to begin 10 hours in bed. (The Ceres Bureau of Health suggests that 9 hours of sleep out of 27 is commonly adequate. For those seeking guidance for the remaining hour, see Dept. of Population Expansion Bulletin 3762B.)

6:00 p.m.: Nighttime sunrise. You are asleep.

9:30 p.m.: Approximate time to begin meditative wakeful state (called "watching" on Earth before artificial lighting was common). [11]

9:59 p.m.: One more minute before it is 1:00 a.m.

1:00 a.m.: Transition into a.m. period. New calendar date starts.

1:30 a.m.: Nighttime sunset. End roughly hour-long "watching" period and go back to sleep.

6:00 a.m.: Sunrise. Jump out of bed again if not already up.

Indoors, one would need to walk in a more controlled fashion, without pushing downward against the floor more during some parts of the average step than other parts, because that would tend to send you flying upwards, potentially making it hard to stop before careening into the nearest wall or hitting your head on the ceiling. To walk the Cerean way, push backwards against the floor without any extra downward push, using your gluteal (buttock) muscles to "pull" your heel backward with each step. The trick is to not push off with the front of your foot. Try it! It feels odd at first here on Earth but, with a just a little practice, you'll be ready for your next vacation… to Ceres!

Extraterrestrial terrorism

Terrorists and gunslingers on Ceres could create a form of chaos not possible on Earth. Bullets could be fired into orbit.[12] When the orbit brings them nearly to the surface, they could actually hit an innocent bystander years after they were fired. There is no atmosphere to slow them down! Earth firearms will generally work on Ceres. Muzzle velocities vary depending on type of gun. Some firearms cannot launch a bullet into Cerean orbit, while others fire a bullet so fast that, instead of orbiting, it would escape the gravity well of Ceres entirely and fly off into space. For intermediate power firearms, the true velocity can be made faster or slower than the muzzle velocity by firing in the direction Ceres rotates, or against it, or some other direction. This is the same principle by which rockets here on Earth are often fired near the equator: The ground is moving eastward (which is why the Sun rises in the east) fastest there because at the equator the Earth's surface is farthest from its axis of rotation. This gives the rockets a speed boost. Thus, fired carelessly or maliciously, many guns could put a bullet into an orbit around Ceres that comes arbitrarily close to the surface, leading to the following future possible news story…

"Recently, a Cerean youth on trek nearly lost her life from an orbiting bullet! Probably it had been fired many years earlier during the pioneer days, when corporate "security" forces from Earth battled over Ceres' valuable mineral deposits, which are now controlled as a public trust by the Bureau of Mines. People talked about it non-stop for weeks (since normally, nothing happens on Ceres). She jumped up only ten feet (3 m), equivalent to jumping up about 3 inches (8 cm) on Earth, and an orbiting bullet punctured her space suit, grazing an elbow. Fortunately other youths in the party responded quickly, heroically patching her space suit before too much air was lost, then using a special tourniquet to temporarily apply pressure to the injury through the suit. While the trek is the bridge to adulthood on Ceres, no one wants such heroism to become necessary. Guns aren't much use on Ceres anyway, except for staple guns.

More exotic colonization options

If Ceres is not exotic enough, let's consider a few even more distant options. What about colonizing the giant planets Jupiter, Saturn, Uranus, and Neptune? And then there is (maybe) Planet Nine. Predicted to lurk far, far away, about 20 times further out from the sun than Neptune, it would be about 10 times the mass of the Earth and thus, probably, like the other giant planets — if it really exists.[14] Theories that a new planet in the solar system awaits our discovery have come and gone over the years, but who knows, maybe they will find this one! The giant planets get little sunlight useful for growing food and providing power, but energy would have to be obtained somehow; perhaps through fusion, a technology yet to be figured out. These planets also have poisonous atmospheres, and don't have a definite surface to build or even float on because the atmosphere just gets denser and denser (though at least warmer but then hotter) as you go down. Colonizing these planets would make big balloon colonies hovering 30 miles (50 km) above the surface of Venus seem like a piece of cake. And yet, the hovering, enclosed bubble colony concept would be the way to do it. Although unrealistic today they could work in principle, and as technology advances and gets cheaper, eventually (or sooner, if the singularity arrives as many hope) such colonies could become feasible.

On the other hand, the giant planets are not without temptation: A few scientists suggest that far enough down, large diamond bergs might drift in liquid carbon, which might tempt some to try to fish them out.[13] A typical iceberg on Earth might easily weigh a couple hundred thousand tons; a diamond berg that heavy would be 907 billion carats plus change (over 180 million carats of "change," to be nit-picky). That's a big diamond and expensive, too. It would look quite impressive set into a cocktail ring. Of course the cocktail party would have to be in a very big hall to accommodate someone wearing a ring with a diamond that huge, and the wearer would have to remove her ring in order to walk around. But at least it would be theft proof, since it's not easy to steal a 200,000 ton diamond.

If you think the task of setting a diamond the weight of an iceberg into a cocktail ring might be a head-scratcher for even the most experienced jeweler, how about an entire planet? Not in our solar system, certainly, but astronomers have reported a planet 4,000 light years away that is likely to be mostly a giant diamond.[15] This planet, Lucy's Diamond, is officially named "PSR J1719-1438 b."

Setting aside the more daring, diamond-struck adventurers, a more likely bet for colonization would be certain moons. Our solar system has no Pandora, as popularized by the movie *Avatar* but first described in the Strugatsky brothers' 1960's Russian sci-fi series *Noon Universe*. But Saturn's moon Titan has rivers, lakes, and rain. True, they are mostly liquid methane, without liquid water, but hey, at least the ground contains lots of rocks made of ice. Jupiter's moons Europa and Ganymede have plenty of ice too, and subsurface liquid water, but also lots of dangerous radiation at the surface. Indeed, considerable ice is available on many moons of the gas giant planets. Jupiter, Saturn, Uranus, and Neptune collectively have no less than 14 moons with diameters of over 500 miles (800 km) and a lot of smaller moons. If Planet Nine exists it probably has moons as well. Some of these moons are more forbidding than others.

Talk about forbidding, it's a cold day in hell on Jupiter's frozen-over moon Io. Io is considered the most volcanically active body in the solar system. Rather than mere ordinary liquid rock, however, large quantities of sulfur issue forth.[16] Sulfur is the modern name for what used to be called "brimstone," and Io is hellish indeed. There are even thought to be fuming lakes of the stuff on Io. Yet most of the sulfur (er, brimstone) covered surface is frigidly cold.

At least the intense radiation bathing the surface interacts with Jupiter's magnetic field to produce huge amounts of high-tension electricity... just waiting to be tapped by its future diabolical, energy-hungry, and chilly but doubtless space heater-carrying inhabitants. But who would want to reside in such a living hell? Maybe no one. Perhaps Io could someday be used for incarceration. An implacable prison from which no one could escape, since prisoners would not be issued space ships. Io would put ordinary supermax prisons to shame, being more of a superdupermax. Meanwhile Jupiter hangs in the sky, an enormous, glowing, many-featured orb. It should make quite a beautiful sight (viewed through the required heavy duty radiation shielding, of course).

The outer limits: Pluto and Eris

Pluto gets no respect. It was discovered by self-made astronomer Clyde Tombaugh on a Tuesday, February 18, 1930, in Flagstaff, Arizona. Last hired and first fired of the planets, the International

Astronomical Union de-listed it in 2006, pulling its planetary operating license and forcing Pluto into retirement after a mere 76 years on the job. Technically renamed "134340 Pluto," this formerly proud body is now forced to moonlight as a "dwarf planet" to the continuing consternation of Plutophiles everywhere. To add insult to injury, it is not even the biggest dwarf planet. That honor currently goes to Eris, discovered in 2005 and, frankly, not that well-respected either: Many people have never even heard of it.

As targets for colonization Pluto and Eris have problems, though nothing like those associated with the gas giant planets or even Venus. The main problems are getting there in reasonable time, obtaining enough energy to warm the colony (which will be sealed to keep the air in), growing food or manufacturing nutrients, and generating electricity.

The New Horizons spacecraft launch of January 19, 2006, destination Pluto, was designed with a planned travel time of 9 years and 176 days. However a one-way trip to colonize Pluto would take even longer owing to the extra time needed to slow down in order to land. Eris is a bit less than four times as far away as Pluto. Sometimes it can actually be closer to the Sun than Pluto, but that won't happen for about another 800 years.

Prospective colonists will have severe energy challenges once they manage to get there. Pluto's distance from the Sun ranges from 29.7 times Earth's average distance, up to 49.3 times Earth's, depending on where it is in its rather uncircular orbit. For Eris, the distance ranges from 37.8 to 97.6 times Earth's. Unfortunately the brightness of the Sun is related to the square of the distance, not the distance itself, so the Sun on Pluto is actually between 880 (or 29.7×29.7) and 2,431 (or 49.3×49.3) times weaker than on Earth. That's actually plenty of light to see around in, equivalent to shortly after sunset on Earth.[17]

With the Sun so weak, sunburn would be the least of your worries. In essence, you'd need 2,431 computer-controlled mirrors all reflecting the Sun to the same spot, to be sure to get up to at least Earth's sunlight intensity at that spot. Then, if that spot was inside a transparent, airtight bubble, you could grow crops there, right at that spot. If you wanted to grow 1 acre (0.4 hectare) of crops, on the order of what's needed to support a person, this simple analysis shows you would need up to 2,431 acres (984 hectares) of computer-controlled mirrors. Fortunately, in practice somewhat fewer mirrors would be adequate because the Sun shines on Pluto's surface without attenuation, while Earth's atmosphere blocks some sunlight. Also, during favorable segments of the orbit you'd need less, as "few" as 880 acres (356 hectares), but you do have to eat during the unfavorable times too. For Eris, the sunlight gets as low as 9,518 times weaker than here on Earth (implying an upper bound of 9,518 mirrors for Earth-style light intensity). Though this sounds dim, it is actually about 35 times brighter than the full Moon, so you could see well enough to get around without artificial light or mirrors. Still, mirror manufacturing definitely needs to get more cost-effective before growing plants there becomes feasible. Perhaps some other energy source can be found besides the distant Sun. Or perhaps it would be more feasible to use the scarce solar power in a more efficient way by synthesizing nutrients artificially in a chemical laboratory, an idea that follows the circa 1894 proposals of French chemist Marcellin Berthelot.[18] Once the energy problems are solved — bon voyage!

Beyond the outer limits: Colonizing the cosmos

Perhaps the biggest obstacle to colonizing our galaxy, the Milky Way is the huge distances between stars. The nearest star to Earth is Proxima Centauri, 4.24 light years distant. Traveling in a normal spaceship at what is currently a realistic speed of 50,000 km/hr, it would take about 91,000 years to get there. This makes the trip seem unlikely without radical advances in spacecraft speed technology. As for freezing people for 91,000 years and thawing them out afterwards, good luck, you'll need it! Oh yeah — and once you get there, we have no evidence yet that there are any planetary bodies orbiting Proxima Centauri on which to land. So is there any way to save the concept of interstellar colonization? Perhaps.

Mysterious nomad planets

Although stars are far apart in our neighborhood of the Milky Way, rogue, or nomad planets, *might* be much closer. In fact, there might be 100,000 times as many such nomads floating around in space as there are stars.[19] We just haven't seen them yet. After all, nomad planets don't shine like stars so they are hard to detect and thus only a relatively small number have been detected so far. These have catchy names like S Ori 52, UGPS J072227.51-054031.2, and CFBDSIR2149-0403. Recent advances, however, give hope that they are out there waiting to be found by the thousands as detection technologies improve.

Nomadic planets have two particularly neat characteristics. One is that, because there are so many of them, the nearest ones are a lot closer (hence easier to get to) than the nearest stars. Another is that, unlike a star, it is possible to land on many of them (some will be gas giants and lack surfaces to land on, but even those may have landable moons).

Once we land, a major problem is that, since these nomads are so far from any star, their surfaces are exceedingly cold. One potential solution is drilling technology that would enable "mining" heat from deep under the surfaces of these interstellar nomads. This geothermal energy could be used to generate electricity and other forms of energy needed to manufacture nutrients and otherwise sustain human life, however dreary living underground and eating artificial food might be (soylent gray, anyone?). Such drilling technology is clearly within our grasp on Earth, although exporting it to another planet just as clearly requires some advances. Perhaps our soylent gray-eating colonists could just dig relentlessly by hand? No one claims that space colonization will be easy!

Another energy possibility is controlled fusion power. In principle, it's a great solution — the Sun uses it so why can't we? The problem in practice is the extreme technical difficulty involved in making it work even here on Earth. Despite these challenges, an important milestone was reached in 2014 when the US National Ignition Facility reported getting more energy from fusion than was needed as input to make the fusion happen.[20] Still, digging a deep hole is surely easier.

How close are these mysterious nomad planets? There are 9 stars within 10 light years of Earth. Therefore at the guesstimated 100,000 nomads per visible star, there would be about 900,000 nomad planets within 10 light years. That suggests 900 nomads within 1 light year (since a sphere of space 1 light year in radius has 1/1000th the volume of a sphere 10 light years in radius, just like a cube of space 1 light year on a side has 1/1000th the volume of one 10 light years on a side). By the same reasoning, the nearest nomad planet could easily be just 1/10 of a light year away, a lot closer than Proxima Centauri. How long would it take to get there?

The likely distance in astronomical units (one AU is approximately the average distance from the Earth to the Sun) is about 6,600 AU, or just under a trillion miles (1.6 trillion km). At a speed achievable with today's spacecraft of 50,000 km/hr, it would take about 2,200 years to get there. This is bad, though not as bad as the 91,000 years to get to Proxima Centauri, and at least there would likely be a place to land. If we can just get travel speeds up by a factor of 100, it would be a 22-year trip. It would be easier to hopscotch through the galaxy, colonizing nomad planets about 6,600 AU apart, than it would be to jump from star to star, as stars are so much farther apart.

This hopscotching process has an additional advantage. It doubles as a way for humankind to get to the stars. And many stars have Earth-size planets in what is called their habitable zones (HZs). Specifically, about 22% of sun-like stars plus about 15% of red dwarfs, which are also friendly, have HZ planets, totaling an estimated 40 billion such planets in the Milky Way galaxy.[21] Wikipedia maintains a list of the best such planets that are known (en.wikipedia.org/wiki/List_of_potentially_habitable_exoplanets). While 40 billion is a lot less than the likely number of nomad planets, it is still several HZ planets per human being!

What we can do now

Tracking the advance of space technology

It would be useful to understand and track the technical details about how, and how quickly, space-faring technology is advancing. Research on elaborating, testing, standardizing, and using such technology tracking methodologies could be better supported by academic research and government research funding. The current leading approach could be expanded upon. It is termed "Technology Readiness Levels," or TRLs, and is used in the US by the National Astronautics and Space Administration (NASA) and the Department of Defense (DoD) as well as other organizations worldwide. TRLs classify relevant technology levels such as "speculative" or "mature" on a spectrum. "Speculative" describes, for example, proposals for faster-than-light travel via cosmic wormholes. "Mature," on the other hand, could be applied to space systems that reach operational status, like the US space shuttles of the early 21st century.

Proposals for a space elevator are somewhere in the middle. A space elevator is a long, long rope attached to the Earth (say) at one end and stretching out into space, with a weight on the far end to pull it taut as the Earth's rotation swings it round and round. It's just a giant version of you whirling a string with a weight on it around your head. To get something out of a gravity well much more cheaply than using a rocket, just have it climb the rope into space.

From sunbathing to moonbathing to starbathing

Here on Earth, it is useful to keep in mind that moonlight is hundreds of thousands of times dimmer than sunlight. This means that, though sunbathing is hazardous even with sunscreen, moonbathing is perfectly harmless and perhaps even fun. Feel free to go right ahead. But don't expect to get a moontan as the light is simply too muted and pale, considerably less than even sunshine on Pluto or Eris. So that's the situation with moonlight... but what about starlight?

The brightest star in the sky is Sirius, with an apparent magnitude of −1.47. This is quite a bit dimmer even than the full Moon, whose apparent magnitude is about −12.9 (the lower the apparent magnitude, the brighter the object). The Sun has an apparent magnitude of −26.7, so the difference in apparent magnitude between the Sun and Sirius is just over 25. A difference in apparent magnitude of 5 is defined as a 100-fold difference in brightness, so 25 levels means a 100-fold brightness difference which has been compounded 5 times, or $100 \times 100 \times 100 \times 100 \times 100$, i.e. 10 billion. Thus starlight from Sirius would need to be concentrated about 10 billion times to reach the intensity of sunlight. At roughly 4 billion square inches per square mile, that's about 3 square miles devoted to focusing starlight from Sirius onto a single square inch (or 1.2 km^2/cm^2). Since starbathing requires more than a square inch of light, that pretty much means an entire metropolitan area or its equivalent devoted to focusing Sirian starlight directly onto your beach towel. Anyone hoping to starbathe their way to a tan should just forget it. Still, starbathing at night using ordinary, unconcentrated starlight is perfectly fine, just like moonbathing. Enjoy it to your hearts content.

Communing with your inner colonist

It's easy to check out NASA's series of posters promoting tourism to various extraterrestrial bodies — just visit www.jpl.nasa.gov/visions-of-the-future.

Why not take up vegetable gardening, just like a extraterrestrial colonist? Food production may involve growing plants, just like on Earth. Off-Earth farming might resemble vegetable gardening more than commercial agriculture. Instead of acre after acre of a single crop, colonists will grow a number of different kinds of plants in modest quantities inside the colony's cavern or bubble. That will give the colonies more diverse and thus robust ecosystems. It will also make for a more varied diet for the colonists, which is healthier as well as better-tasting. Your gardening experiences here on Earth, both good and bad, will mirror to a significant degree those of future space colonists.

Non-crop food production

Growing crops in an extraterrestrial location is risky. The difficulty of maintaining the rather narrow range of temperature, light and moisture conditions necessary to sustain them makes crop failures and even crop species extinctions a serious existential risk to the colony. A much more reliable approach is to manufacture nutrients chemically. Although this enhances the survival prospects of a colony, it is a lot less enjoyable for the colonists who, instead of spending pleasant days tending growing plants in spacious (and thus difficult to construct) surroundings, must spend their lives building and maintaining chemical manufacturing processes in potentially cramped, dim and perhaps hot or cold abodes. That said, we should prepare for this now, here on Earth. Although it is hard to design and rehearse methods of growing crops on, say, Mars, here on Earth, nutrient manufacture on Mars or anywhere else can be much more easily studied on Earth.

This task starts with a list of every essential human nutrient. Each nutrient on that list needs to be matched with a manufacturing process for creating it in a form suitable for human consumption. Whether it is a simple trace element like iron, an organic substance like an essential amino acid, a lipid, carbohydrates, or a complex molecule such as one of various vitamins, a suitable manufacturing process is unique for each one. We can design those processes here on Earth for use anywhere in the universe that suitable energy and simple chemical elements or other raw materials can be obtained. Moreover, the expense and difficulty of developing a manufacturing process for a typical essential nutrient, suitable for deployment at some off-Earth location, is reasonable. Since the number of essential nutrients is in the dozens, a complete program for nutrient manufacture, tuned to any given extraterrestrial location, is within reach. Such programs are needed and should be undertaken as soon as possible.

Chapter Fifteen

Tastes Like the Singularity

Some think the world as we know it will soon end, ushering in an unimaginable (but hopefully utopian) future. This chapter explains why it is called the "singularity," and why it's exciting. Will it radically transform the fabric of reality?

If the artificial intelligence singularity happens, the world will soon be under the sway of entities much smarter than ourselves. Then things will be, as J. B. S. Haldane put it in 1928, "not only queerer than we suppose, but queerer than we *can* suppose."[1] We will be no more able to understand, outwit, or control an entity much smarter than ourselves than a cow can a person. At least that's the theory. The counterpoint is the claim that for humans to aspire to build a robot more intelligent than ourselves is impossibly absurd, like for a monkey to aspire to reach the Moon by climbing a tree. But who is to say that no monkey has ever tried?

Singularities

A singularity is a particularly dramatic type of situation in which a mathematical description stops working. For example, suppose we describe some unknown quantity x using the equation $x/2=5$. Then $x=10$, because then $10/2=5$. No singularity or anything else unusual there. And if the '2' decreases, x gets bigger. But what if it decreases to 0? Then we have $x/0=5$, and there is no solution for x because ordinary arithmetic does not say what happens if you divide by zero. The value of x in this case is said to be "undefined." More generally, when what you are dividing by becomes zero, whether it is money on line 54 of the infamous Connecticut 2008 income tax form CT-1040,[2] volume of the mass at the center of a black hole, or whatever it may be, you've encountered a singularity, and there is no answer. Connecticut tax authorities might take a dim view of the matter, but astrophysicists are concerned: It is thought that inside a black hole, the gravitational field forces everything inside the event horizon into a dot at the center. Since calculating density requires dividing by volume (density = mass/volume), the density of matter at the center of a black hole would be undefined if the volume truly became zero, thus creating a singularity. Luckily for reality, physics has proposed theories, like quantum gravity, that avoid this mathematical modeling problem by allowing volume to get very small while preventing it from becoming precisely zero.

The AI singularity

Somewhat similarly, the AI singularity occurs when the attempt to calculate the limits of computer intelligence breaks down, seemingly predicting an unending spiral toward infinite intelligence. However, truly infinite intelligence can't happen anymore than a misguided Connecticut resident could have made the entire state financial infrastructure go "poof!" in 2008 by trying to fill out line 53 without having any Connecticut adjusted gross income. Similarly, there is something very strange at the center of a black hole, but it is real even if we don't yet know exactly what it is.

Singularities are properties of defective descriptions of the real phenomena, not of real phenomena themselves. For the AI singularity, the real phenomenon involves computers getting steadily more powerful. They will come to outpace human intelligence in more and more ways. Computers have long exceeded our intelligence in speed and reliability of arithmetic calculation. They can play chess better. They can play *Jeopardy!* better. Each new generation of modern computers can only be designed with the help of previous generations of computers. This process will continue but there will never be a moment when computers suddenly become smarter than humans and take over, because intelligence is seemingly so complex and indefinable a concept that no single satisfactory measure of it exists, or perhaps can exist, and therefore there can be no clean line of demarcation between less intelligent, and more intelligent, than humans.

Thus we won't wake up one day to find our previously loved machines suddenly informing us, as the notorious Japanese video game *Zero Wing* put it, "All your base are belong to us."[3] But the trends do suggest that they are gaining greater and greater intelligence and influence on our lives, perhaps eventually with revolutionary results.

So what is intelligence and how can we tell if computers have it? The tricky question of properly defining and measuring intelligence does not seem to be solvable. At least, it hasn't been solved yet. Nonetheless, it is obvious that intelligence exists and that some people have more of it than others. The classical approach to defining when computers have intelligence is the so-called Turing Test, created by British code breaker and war hero Alan Turing.[4] (Turing was long thought to have later committed suicide by eating a poisoned apple, like Snow White, after being convicted of homosexuality and then "treated" with hormone injections in accordance with the British legal process of the time. However the detailed cause of death is now in dispute.)

In essence the Turing Test says that, in a keyboard chat session, if one can't tell whether one is texting with a chatbot or a person, and it is a chatbot, the chatbot should be considered intelligent. This is a clever idea, though not perfect:

- One problem is its assumption that writing intelligent-seeming text messages actually requires intelligence. Maybe it doesn't.
- Another is that people must be able to tell the difference between text messages produced by intelligent vs. non-intelligent entities. Maybe they can't. For example, in 2014 a

chatbot called "Eugene Goostman" posed as a 13-year old speaker of English as a second language and, many believe, passed the Turing Test. However no one seriously claims the bot is actually intelligent.

- A third is that it ignores the possibility that a computer could be intelligent yet still unable to pass the test, somewhat like a person not fluent in your language, though intelligent, would be unable to pretend fluency.

Turing Test considered as a red herring

The first chatbot was the 1967 program ELIZA.[5] J. Weizenbaum, its creator, wrote, "ELIZA created the most remarkable illusion of having 'understood' in the minds of the many people who conversed with it."[6] ELIZA is probably too primitive to have that effect on today's much more sophisticated computer users and does not pass the Turing Test (it's been tried). Yet the Turing Test is useful and has inspired a regular contest. Since 1991 the "Loebner Prize"[7] has been awarded yearly to the owner of the chatbot contestant that comes closest to fooling a panel of human judges. As a side note, AI pioneer Marvin Minsky is on record as offering a $100 cash prize to anyone who can get Loebner to stop sponsoring the "stupid ... obnoxious and unproductive" prize.[8] For his part, Loebner (a single gentleman and advocate of legalized prostitution) argues this actually makes Minsky a co-sponsor of the prize, since he would have to give his cash offering to the owner of the first chatbot to fully pass the test, finally winning Loebner's Grand Prize and thereby ending the annual competitions.

The Turing test is clearly suspect on logical grounds alone (as explained above), and most anyone working on chatbots will confirm that, in practice, they don't consider their impressive creations to be truly intelligent. But that is likely to change at some point. Chatbot performance appears to be generally improving from year to year, so progress is occurring. Indeed, the winner's performance in the

Loebner Prize competitions over time would appear to be one way to measure progress in computer intelligence. Although not a perfect metric, it is an interesting one.

Other metrics exist as well, also imperfect but very different from chatbot performance and from each other. One measures a computer's creativity.[9] Game playing is another fruitful source of potential ways to measure improvements in computer abilities that seem to require intelligence, because games tend to provide a clear context that supports quantifying performance. Progressively increasing computer chess performance had already won the world championship years ago, in 1997.[10] Soccer is different from chess, but robots compete in soccer in the robocup games, held yearly since 1997. Their soccer performance represents another metric for intelligence.

A trajectory of improvement in a composite of different tasks indicative of computer intelligence is more convincing than one of improvement in any one metric, in part because intelligence itself is such a complex, composite attribute. A useful approach might be to keep a running count of human games that machines are able to play better than humans. Chess and *Jeopardy!* are already there, but soccer is not. (Hopeful robocup organizers however have a goal to, "By the year 2050, develop a team of fully autonomous humanoid robots that can win against the human world soccer champion team."[11])

What we can do

The AI singularity will not rear up overnight, instantly changing your life dramatically for either the worse or, as riskily assumed by some, the better. Every age has its messianic movements and its rapturous apocalypticists. Still, AI does appear to be improving. Computers are already far in advance of human arithmetic intelligence, and society has leveraged that into many benefits, from calculators to income tax software to spacecraft navigation systems and more. This will continue to happen with other computer capabilities. Thus the number of such capabilities that exceed human performance, such as mishap-free motor vehicles, will grow progressively.

Popular movies have long relied on the concept of a secretive and misanthropic "mad scientist" who creates a robot of great capabilities. That will probably not happen. It takes a sizeable community of skilled humans to create even a pencil. Referring to everything from chopping trees for the body to making rubber and metal for the eraser, L. E. Read notes in his classic essay *I, pencil*, "... not a single person on the face of this Earth knows how to make me."[12] Even the simplest computer is obviously far more complex than a pencil. For a robot to create another robot of greater capability than itself would require either large numbers of humans and other computers to help, just as it does now, or a single robot with the intelligence, motor skills, and financial resources of thousands of humans and their computers, factories, banks, etc. How many thousands? There is no way to know for sure. But consider that human societies of thousands, once isolated, have lost even basic pre-industrial technologies.[13] Tasmania is a well-known example.

An important need is for metrics that can tell us, in practical terms, the rate of progress by which artificial intelligence is marinating society. Arithmetic, the Turing Test, chess, *Jeopardy!*, soccer, and even self-driving cars are interesting but do not fit the bill by themselves. As factors in a richer, composite metric, however they can play a part. Another factor that might be useful is to count the rate of new AI applications becoming available over time. A suitable composite metric should be debated and converged upon by society.

Chapter Sixteen

Chasing the Future: Spoilsports of the Prediction Game

What are the seven wet blankets of the prediction endeavor? They include the butterfly effect, the uncertainty principle, the observer effect, and others. It means we can't predict the future with certainty. But we can try.

Ever have a day when everything went wrong? You predicted you would have a normal day, but your alarm clock didn't ring. Already running late, you couldn't find your briefcase or backpack. Staggering to your car, it wouldn't start. Later, you found out you missed a surprise meeting or maybe a college quiz. It's not you, it's the prediction game. It gets worse as the prediction horizon gets farther into the future. For example, how could anyone a thousand years ago have predicted the ongoing competition between paper printing and computer screens? Neither one had been invented yet! The same type of problems apply today when we contemplate human society thousand years from now. In that spirit, here are seven spoilsports of prediction — the wet blankets that say why we just don't know for sure what is going to happen, no matter how much we want to.

Wet blanket #1 — the observer effect

To figure out what happens next, you need to know the state of things now. For example, if you hit a billiard ball, you can't predict what will happen next without seeing the current layout of the pool table. Unfortunately, as a matter of principle, the observer effect holds that the act of observing the current state of the system, changes it. Technically, it "perturbs" the system. In physics this is most noticeable for very small things or faint effects. In principle, it applies to any scenario.

Light pushes on the illuminated thing — the push is slight, yet real enough that sunlight-driven solar sails have been suggested as a propulsion method for space travel. Electrical measurements affect the electricity. Asking someone something about themselves changes them; asking someone what they think of you changes the way they think of you. The observer effect has filtered into popular culture as well; it's even the name of a *Star Trek: Enterprise* episode.

The list goes on. If you are watching over kids, they act differently. So do adults. Publicizing economic indicators affects the economy. Political polls affect voting behavior. Stock market prices are more than economic signals — they affect themselves, causing bubbles and crashes. Perhaps game theory could be subtitled, "Using the observer effect to win." The social competition theory of human brain evolution even holds that game theory is the key to why we are intelligent.[1] We're watching us.

Do you think the observer effect can be avoided? That we could predict the future of anything, at least in concept, if we could just observe the position, velocity, and mass of every subatomic particle involved, crunch on a computer to figure out what bounces off of what and where everything ends up, then zoom out and, *voilà*, that is the future? Alas, that just brings us to…

Wet blanket #2 — the Heisenberg uncertainty principle

You could not play pool without seeing where the billiard balls are on the table. The real world is analogous: The uncertainty principle is physics' way of explaining that it is impossible to reliably predict the future of a complex real system without knowing the exact positions and momentums of its components; and (the kicker) you can't know those things exactly, so you can't reliably predict the future.

In the realm of subatomic particles, for example, the more accurately you know position of a particle the less accurately you can know its momentum and vice versa. That's just how things work. A fun 3-minute video introduces the Heisenberg uncertainty principle at www.youtube.com/watch?v=iFwRAvpWDB8. The principle applies in theory to all objects, not just electrons (which would be bad enough, since electrons are part of all normal matter). For example, it applies to photons, the pieces of light, as well. See the experiment in the video at www.youtube.com/watch?v=KT7xJ0tjB4A.

You can do that experiment at home. Get a laser pointer and an opaque sheet of something you can cut a narrow V-shaped notch in. I used a black vinyl notebook cover and used an ordinary sharp scissors to cut a narrow V a few inches (several cm) long and an inch (couple cm) wide at the widest point. In a dim room, shine the laser beam through the V at a wall. Move the beam close enough to the point of the V, and the bright dot on the wall will smear — the Heisenberg uncertainty principle in action.

The uncertainty principle connects uncertainties about position and momentum mathematically. Momentum is the amount and direction of "shove" that an object could impart: A slow-moving but heavy bowling ball might have as much "shove" as a much faster-moving but lighter bullet, and if you shot an approaching bowling ball head-on and they both had the same amount of momentum, they would exactly stop each other. The uncertainty in an object's momentum, times the uncertainty in its position at a particular instant, can never be zero. In fact, it must be at least as large as a very small number close to "Planck's constant." But small as it is, it still means that you can't be certain (that is, have zero uncertainty) about any subatomic particle or indeed any other entity's position and momentum. Why? Because then it would be the case that uncertainty in position would be zero, and uncertainty in momentum would be zero, so multiplying them to get their product yields zero ($0 \times 0 = 0$), yet that product cannot be zero since it must be at least that tiny number related to Planck's constant. Worse, you can't even have zero uncertainty about just the position because then

$$\text{position} \times \text{momentum} = 0 \times \text{momentum} = 0,$$

and that zero result is forbidden by the uncertainty principle. By the same reasoning, one can't have no uncertainty (i.e. full certainty) about just the momentum, either.

To summarize: Fully describing a system with the exact position and momentum of everything in it is required to predict its future. But these properties are in principle impossible to get with full accuracy, because the uncertainty principle forbids it.

So if the observer effect doesn't stop our prediction ambitions, the uncertainty principle will. But what if we could control both at least enough to predict with some degree of confidence? Alas, we're still not out of the woods, because next we run into the esoteric physics phenomenon called "quantum tunneling"...

Wet blanket #3 — quantum tunneling

We intuitively think that objects are localized in space, but quantum physics says otherwise. In fact, objects have wave-like characteristics and are actually "smeared" over a space within which they may be said to exist at each point with some probability. A tiny object like a subatomic particle, if near enough to a thin barrier, thus has a certain probability of being on the other side of the barrier owing to this smearing effect. If it turns out to be on the other side, it has thus "tunneled" through the barrier without making a hole in it. This is quantum tunneling (see video at www.youtube.com/watch?v=6LKjJT7gh9s).

Quantum tunneling also applies to the ability of objects to "tunnel" through other kinds of barriers besides solid ones. For example, consider the rather notorious example of an idealized pencil balanced on its tip.

If the tip is sharp except for a tiny flat spot (say, a couple of atoms wide) it would be difficult to balance, but one might think that with sufficient care it could be done. But not quite. Because the pencil is actually "smeared" a little bit, it actually has a rather small probability of being tipped enough to lose balance and fall. Since the smearing is symmetric, it could fall in any direction. The probability of being tipped enough to lose balance is small enough that a single such pencil would be unlikely to fall for a long time.[2] But get enough pencils together and one will fall soon enough. For example, balance an array of 1000×1000 of these pencils and one will fall, knocking over others and leading to a domino-like conflagration with an average (but unpredictable) delay of around a month. What pencil will start the general crash and in what direction it falls are impossible to predict.

But maybe we're interested in predicting the future of something not so finely tuned. If we can handle the observer effect, the uncertainty principle, and quantum tunneling well enough, could we adequately predict the future of our system? Not necessarily. Our troubles are still not over, because of the "butterfly effect"...

Wet blanket #4 — the butterfly effect

In the movie *The Butterfly Effect*, Evan is a college psych major consumed by guilt over his failure to keep a childhood promise to Kayleigh. He blames himself for her eventual suicide years later. Discovering a way to go back in time and relive events in the past, he resolves to engineer a better present and future — and save Kayleigh — by reliving past events while changing his actions during those events so that they turn out differently.

Who wouldn't do a few things differently given a second chance? Unfortunately he finds that while it is possible to change the past, it is not possible to predict how those changes will play out over time. He succeeds in saving Kayleigh, sort of, but in the new and changed present winds up in prison himself. So naturally he tries again (who wouldn't?). But the results are worse so he tries yet again... and again. Each time there are radically different unanticipated consequences ranging from bad to worse, until finally he lands in an awful last-ditch present that provides only one opportunity to go back for another do-over. I won't give a spoiler, but it's a compelling concept, which might explain why the movie was a box office hit despite being panned by the critics for its defects.

What is the butterfly effect?

A butterfly flapping its wings will create a small atmospheric disturbance. That disturbance will propagate unpredictably. Months, perhaps years later, a hurricane may track in your direction instead of mine — because of those tiny flaps.

Models of certain atmospheric cycles are indeed known to depend unpredictably on seemingly trivial present events. Special water wheels have been built to illustrate this process. Video clips of some of them are at

www.youtube.com/watch?v=zhOBibeW5J0,

www.youtube.com/watch?v=VumQmC2jJbU, and

video.google.com/videoplay?docid=-355587954903008142.

In the words of the butterfly effect's discoverer, meteorologist Edward N. Lorenz (1917–2008), "... prediction of the sufficiently distant future is impossible by any method, unless the present conditions are known exactly. In view of the inevitable inaccuracy and incompleteness of weather observations, precise very-long-range forecasting would seem to be non-existent."

As goes weather prediction, so goes prediction in other topics. Such questions may plausibly include, "How long will *Homo sapiens* be the dominant species on planet Earth?" "How will the average human lifespan change, and when?" Many others range from personal to planet-wide, from politics to science, technology, and of course economics.

But suppose, using sufficiently precise measurements and computers, you could control the butterfly effect? You can't, but even if you could, you still have not solved the prediction game, because of external perturbations…

Wet blanket #5 — external perturbations

To figure out what happens next, you need to know where things are now. But you also need to know what outside influences will impinge on the system between "now" and "next." Those influences can affect the evolution of the system. That's why they're called "influences."

Imagine for example the Lorenz water wheel (or watch one: www.youtube.com/watch?v=zhOBibeW5J0). Given a steady, predictable stream of input water, it still spins forward and backward unpredictably because of the butterfly effect. But it gets worse if it is spinning in the rain. Now, the input is no longer steady and predictable. Every raindrop is like another butterfly whose tiny effects change the direction of the wheel unpredictably at some future time. More generally, any external nudge to a system is like that butterfly.

Let's identify some external influences likely to affect the future course of systems of interest. If the system is a nuclear family whose dynamics we understand (let us riskily assume), any attempt to predict it in detail will soon founder on the realities of day-to-day events that perturb its state. If a troublesome child commits nuisance vandalism, say, whether he is caught or not and by whom (parents? victim? police?) will likely have some impact on the evolution of the family. Moving up in scale, if the system is a country, the results of the next election depend on much more than the political dynamics. They also hinge on unpredicted international events. Even if the reaction of the country to each event could be predicted, whether and which events will occur is unpredictable from the dynamics of the system itself. What about human affairs generally? There, unpredictability from external perturbations results from everything from actual butterflies flapping their wings, to volcanoes going off, to solar storms, and on and on and on.

Other systems besides Lorenz water wheels and human dynamics have the same problem, because every system is subject to external perturbations, save one. That exception is the entire universe in every detail. You can't simulate that, because no computer is big enough. Even if you thought you had one that was big enough, it would be part of the universe and it can't simulate itself. Also the universe generates its own perturbations that even in principle are unpredictable because of quantum tunneling, like when and in which direction a pencil perfectly balanced on its point will fall. Since those perturbations are impossible to predict, they are effectively external perturbations.

But suppose you could control external perturbations enough, along with all the other wet blankets discussed so far? You've not won the prediction game yet, because until predicting is shown to be worth the effort, why do it?

Wet blanket #6 — why predict if it doesn't matter?

"Solomon Grundy; Born on a Monday; Christened on Tuesday; Married on Wednesday; Ill on Thursday; Worse on Friday; Died on Saturday; Buried on Sunday; That was the end; Of Solomon Grundy." (Mother Goose)

Is that it? Is there no point to life? Because then, the argument goes, there is no real point to the future and thus none to predicting it either. If you find this perspective a downer, you're not alone. Existential depression is said to be a risk among gifted children, for example, so it can start early. Existential nihilism — the distressing feeling "that the world lacks meaning or purpose" is the cause.[3] The problem is the distress, not the laws of the universe, since meaning and purpose are mental and cultural, not physical phenomena.

It is easy to ask if the future matters and conclude that it does not. It is likewise easy to argue that the existence of humanity itself is no great gift to the Earth. In that case, why worry about the future of the human race? Even if humanity avoids destroying itself, in 10 million or 100 million years our descendants will be more or less as different from us as our ancestors were 10 million or 100 million years ago — i.e., not really people. Thus, one way or another, our species, *Homo sapiens*, is most likely destined to end.

So what to do? Some turn to religion for meaning. There is little evidence to objectively justify the metaphysical correctness of any one religion over all the others, except in the minds of some of their adherents. Multiple religions that each believes all the others are wrong presents a bit of a conundrum, if you think about it. But plenty of people fail to understand that. So we're back to the future and its meaning. If the future is ultimately pointless, then why not just "Eat dessert first."[4]? "Eat, drink and be merry, for tomorrow we shall die."[5]? Or "Don't worry, be happy."[6]?

In fact, that is essentially what people do.[7] Business decisions focus on short-term payback, with "long-term planning" designating horizons as short as three years out. Political decisions focus on short-term, narrow-minded goals, a big reason why they are so often inefficient or worse. Of course, it's not just politicians and businesspeople who focus on the short term. Many ordinary people do as well. In that, we heed our roots: Animals are usually short-term in their behaviors.

By failing to predict the future, we act and react only in the short term. However there is at least one good reason to do just that. The farther out we go in predicting, the less likely we will be right, and thus the higher the risk that the effort spent preparing will end up wasted. That is the overarching theme of all of the wet blankets of the prediction game. As Winston Churchill said, "It is a mistake to try to look too far ahead. The chain of destiny can only be grasped one link at a time."

On the other hand, blindly focusing on the near term is a kind of tunnel vision. The risk there is failing to prepare as the future silently approaches... then suddenly jumps up and bites you (or society) on the butt. We can choose to try to anticipate the future, yet all too often, when "societies choose to fail"[8] (J. Diamond) it is because they chose incorrectly when there is no second chance. Compensating for the turbulence of short-term action does not mean weighting distant future events as much or more than immediate ones. Rather, it means weighting them deliberately and appropriately. The weights of future events should be based on their importance if they occur, the likelihood they will occur, the value of procrastinating, and the cost of acting.

Making the future matter

Happy and distressed states exist in humans (and many animals). An intrinsic property of such states is that they matter. When as a child you fell and scraped a knee, it hurt. The hurting is not the root of the problem, nor is the pavement. The knee-scrape is. When the scrape heals, it no longer hurts.

Similarly, one can have a "scrape" on the mind, causing the hurtful feeling that life is meaningless. That feeling itself is not the root of the problem, nor is the bumper sticker-like slogan that life lacks meaning. The mind-scrape is. Changing the pavement is hard; changing the fundamentals of life is also hard. Healing the scrape itself can be a smart solution.

Søren Kierkegaard (1813–1855), Danish philosopher, theologian, and father of existentialism, concluded that both problem and solution lie within the sufferer. The pain of the scrape feels like a problem, but is not the core problem. The pavement, or life, caused the problem but is not the problem. The true problem is the actual scrape, with the solution to be reached accordingly. Ancient philosopher Hillel the Elder identified a path forward: "If I am not for myself, who will be for me? And when I am only for myself, what am I? And if not now, when?"[9]

Wet blanket #7 — the "so what?" horizon

How much is the future of the human race worth? We'll increase it later, but let's start with an admittedly bargain basement $100. If you put $98.04 in the bank at an interest rate of 2% per year, then in a year you'd have $100. That means getting $100 one year from now is only worth having $98.04 now, at least from a "time value of money"[10] perspective. Similarly, getting $100 in 2 years is only worth $96.12 now, because adding 2% to $96.12 gives $98.04 in one year, and compounding by another 2% gives $100 a year later. Extending this reasoning further, the human race in a modest 233 years would be worth just under a dollar now. In 466 years? Less than a penny.

It's safe to say that $100 is an underestimate for the value of the entire human race, at least to us. So let's increase it to a fair (or at least fairer) price. We might multiply the number of people by the value of the life of each and every person on the planet. What is the value of a person's life? Economics (known as the dismal science, even to economists) tells us that the *de facto* value society places on a human life can actually be calculated, and courts of law in fact sometimes do such calculations. Answers vary greatly, of course, but let's say 10 million dollars, to be relatively safe. Multiply that by the number of people in the world and you get a biggish number, $70 quadrillion for the value of the human race.

But wait — maybe you don't trust the financial and legal wizards with something as important as estimating the dollar value of a life. After all, we already trust them with some pretty important things, and they periodically betray that trust, seriously screwing things up for national and even the world economies. Maybe we should use a higher number, just to be more sure we aren't under-valuing ourselves.

How about valuing the human race at a dollar for every single atom in the known universe? At around 10^{80} (1 followed by 80 zeroes, or

100,000,000,000,000,000,000,000,000,000,000,000,000,000
,000,000,000,000,000,000,000,000,000,000,000,000) atoms, that's a lot of cash. Way (way, way) more than the United States has ever printed. There are literally not enough atoms in the known universe to even print that many dollars. Yet, if that is the value humanity's existence will have 9,070 years from now, then the value at present of its existence 9,070 years in the future would be... $100! A scant 466 years after that? Less than a penny. How about the present value, today, of humanity existing in a million years? The answer is a fraction of a penny so tiny that standard spreadsheets, calculators, and computer programming languages can't even state it. They typically just think it is 0, but if you must know, it's actually about $0.000[insert 8,513 more zeroes here]0001.

Wait — someone in the back has a question — yes?

"But it's not just the value *in* year one million we're after. We also need to add in the value in year 1,000,001, year 1,000,002, etc., forever and ever! That's got to add up, eventually." It does. But, it turns out, not very much.[11] The value now is about 100× bigger: less than $0.000[insert only 8,511 zeroes here this time]0001, even at a dollar an atom. The upshot of all this is that there is no logical reason to care whether humanity exists in ten thousand or a million years, at least according to the time value of money approach favored by economists. Therefore there is no need to plan that far into the future, or go to trouble and expense to preserve the Earth indefinitely, or even to bother predicting that far ahead. The time value of money seems indeed to be a wet blanket of the prediction game.

146

Making it personal

Maybe you are still unconvinced. Such sophistry fails to capture the real facts at a common sense, gut level, you say? Then consider the following argument.

You care about yourself, so you don't want humanity to end while you are still alive (it might not be pleasant). You care about your children, or maybe you care about some or even all other children, so you don't want humanity to end during their lifetimes even if you are already gone. You probably even care (or may someday care) about your grandchildren, especially if we assume you will get to know them personally. Furthermore, you care about *their* grandchildren (if maybe a little less) simply out of concern for your own grandchildren, since their well-being depends in part on the well-being of their own future grandchildren. But you have no reason at a gut level to care about the generations after that, because neither you, nor anyone you care about will ever know them, if indeed they ever even come to exist. How much does your heart go out to those who won't exist until long, long after you are not only gone but forgotten too? To put it another way, how much do you care about your grandparents' grandparents — and how much did they care about you? Do you still care in some more abstract, dispassionate sense? Then see the previous paragraph about the decimals with 8,000+ zeroes.

Maybe you are a fast enough breeder, and long enough liver, that you'll care about your great grandchildren and theirs, instead of just your grandchildren. Yet that is still only 6 generations into the future, less than even the biblical 7; a couple of centuries or so at the most. So relax, quit worrying, "eat dessert first."... In particular, don't bother with predicting past the 2-century "care horizon," because there's little point to it. The care (or "so what?") horizon is, thus, our last wet blanket of the prediction game.

What you can do

Don't worry, be happy? Eat dessert first? Distrust the economists? Forget about the future? Pick the one true religion out of the sorting hat, the others being false? If you figure out something new, let us know.

Chapter Seventeen

Warm, Poison Planet

Hydrogen sulfide could poison the oceans and the atmosphere, make the air smell like rotten eggs and turn the sky green. This actually happened, evidence suggests, long ago.[1]

How bad could global warming get, worst case? The planet Venus is interesting in this regard because, despite the fact that Mercury is closer to the Sun, Venus is nevertheless hotter. This is due to Venus's extreme greenhouse effect caused by its very dense, carbon dioxide-based atmosphere. If that happened on Earth, our fair planet would be hotter than a self-cleaning oven. Luckily there seems little reason to fear such a runaway greenhouse effect on Earth. Aside from the fact that it has never happened here before, the Earth may simply not have enough solar energy and greenhouse gas (such as carbon dioxide and methane) to start the runaway positive feedback process that happened on Venus.

Is there any other apocalyptic global warming scenario to worry about? How about a stinking poison released into the atmosphere and oceans by an overheated Earth that not only wrinkles noses worldwide but kills off many living things as well? Welcome to the gray and dead (but warm and balmy) plains, the oxygen-starved waters, green skies, and repellent smell of hydrogen sulfide-poisoned Earth. This is possible. In fact there is evidence it actually happened, long ago.

Hydrogen sulfide, H_2S, is a gas. Chemically similar to H_2O (water) but with a sulfur atom in place of water's oxygen, it is not healthful and wholesome, like water, but very poisonous. A tiny 1 part per million (1 ppm) in the air is easily detectable as the odor of rotten eggs; 10 ppm is a typical occupational exposure limit; 1 part in 1000 in air can cause rapid death. As a youth, I grew algae in several plastic milk jugs of verdantly green water on the window sill, fertilizing occasionally with vegetable peels and such. It worked great, but there was one slight problem: Some vegetables contain substantial amounts of sulfur. This can lead to H_2S dissolved in the water, especially in the muck at the bottom. I finally dumped all the algae water down the toilet rather than move the jugs to another apartment. The process was smellier for some jugs than others, and I ended up with a modest case of hydrogen sulfide poisoning. Main symptom: a parade of colorful, crystalline mental images, presumably the result of hydrogen sulfide-caused inhibition of cellular respiration in the brain. Like humans, most animals and plants are also poisoned by H_2S.

How might H_2S come to poison the Earth? Like it did in the past, according to one mainstream theory. The worst extinction event of all time, much worse than the one that did in the dinosaurs, is believed by many specialists to have been caused by H_2S. This was the much earlier Permian-Triassic (or P-Tr) extinction event of 251.4 million years ago — about 20 million years before any dinosaur was even a gleam in its mother's eye. The vast majority of plant and animal species then in existence went extinct, both in the sea and on land. The P-Tr event is often called the "Great Dying." A similar process could play out in humanity's future, potentially ending it. Here is how.

The causal process begins with a familiar phenomenon, global warming. Massive volcanic eruptions in Siberia are thought to have triggered the warming that caused the great dying. The global warming mechanism may have been the release of carbon dioxide gas from under the Earth's crust into the atmosphere (in contrast to burning fossil fuel, which is doing it now). An alternative (or additional) possible mechanism is that the eruptions released nickel, which nourished a bacterial bloom of Methanosarcina bacteria.[2] As their name suggests these bacteria released methane, in huge quantities. Methane is a greenhouse gas many times more powerful than even carbon dioxide in causing global warming. Warming melts sea ice, which darkens the ocean surface, causing more sunlight to be absorbed and worsening the warming trend. As the oceans warm, any methane hydrate crystals on the deep sea bottom warm too, causing their trapped methane to be released into the atmosphere. The methane hydrate release mechanism is a likely (though still controversial) cause of another global temperature spike called the Paleocene-Eocene Thermal Maximum, 55.8 million years ago, when average global temperatures soared upwards by over 10°F (5.5°C).

As the Arctic warms, the top layers of ocean water there warm too. These layers are thought to be the key to preventing hydrogen sulfide from poisoning the Earth. If they are cold, they do this by sinking. Down to the bottom they go, and when they get there, they move along the bottom, pushed by more waters sinking down behind them. The result is a flow of water, oxygen rich from surface contact with the atmosphere, at the bottom of the ocean. These slow currents travel thousands of miles before resurfacing. They keep the oceans from stagnating and generating hydrogen sulfide, which could end up in the atmosphere.

The process that drives Arctic surface waters down to the bottom, causing the circulation of the worldwide deep ocean currents, is called the thermohaline circulation pump. When it works, it is caused by two major factors.

1) Cold ambient Arctic air cools the surface waters, making them denser and thus tending to sink. This is the "thermo-" part. If global warming prevents the surface waters from effectively cooling, they will not be as dense and thus less apt to sink. This can deactivate the "thermo-" part of the thermohaline circulation pump.

2) As surface waters flow northward to replace the sinking water, gradual evaporation concentrates the salt in these waters, making them denser and thus tending to sink. This is the "-haline" part (from the Greek *hal-*, meaning salt). If global warming decreases the salinity of Arctic surface waters by diluting them, with fresh water from increased rain- and snowfall and from melting glaciers in Greenland and elsewhere, the lowered salt content of the waters will make them less dense and thus less apt to sink. This can deactivate the "-haline" part of the thermohaline circulation pump.

Deactivating the thermohaline circulation pump would devastate the climate of much of Western Europe by ending the flow of warm southern surface water northward toward the Arctic. But it gets worse than that.

Back in the Great Dying, it is hypothesized that after thermohaline circulation stopped (or slowed enough), the oxygen in the deep ocean waters was used up by the organisms that live down there, suffocating most of them. But some microbes don't need oxygen gas dissolved in the water. Some get their oxygen instead from oxygen-containing sulfur compounds. That's bad news, because they then release the villain... hydrogen sulfide (the same noxious stuff that accumulated at the bottoms of those plastic milk jugs). But it gets even worse still.

The hydrogen sulfide slowly accumulated in the ocean waters, poisoning many of the remaining organisms. That explains why the extinction event was so devastating to marine life. Things went from awful to even worse. So much dissolved hydrogen sulfide gas eventually accumulated that it started leaking from the water into the atmosphere. Because only a very low concentration of hydrogen sulfide (H_2S) is needed to create a bad stink, if this happens during the human era the first blatantly obvious sign will be the smell of rotten eggs. It will be pervasive. Though unpleasant, it is not harmful in very small concentrations. As it accumulates in the atmosphere though, the smell will go from bad to worse. Eventually the increasing amount of H_2S will start poisoning land life and make the sky turn green. This process, which can explain the devastation to land life during the Great Dying, could happen again.

Recommendations

There's no need to buy a gas mask to prepare for this just yet. Things won't start getting really bad during our lifetimes. But this could eventually become an existential risk to our species. Thus, scientific study is important. For example, it has been discovered that the Black Sea, below the surface, contains vast quantities of water that is both deoxygenated ("anoxic") and high in hydrogen sulfide. This is exactly the type of water we have been discussing which, if present in oceans worldwide, would pose an existential danger to humanity. Such water is described scientifically as "euxinic," after the ancient Greek name for the Black Sea, Pontos Euxenios.

In general, there is a risk that things we do in our lifetimes may trigger an extinction event later. Obviously it would be the height of irresponsibility to let that happen. Yet there will always be forces of irresponsibility. If such forces succeed their victory would be Pyrrhic indeed.

Chapter Eighteen

Day of Contact

To the reader: This is a story, but it could happen...

The following account was automatically translated, necessarily imperfectly, from information received from the year 2439, via quantum real-time semantic tunneling (QRST). Will the meeting recorded here take place on Earth, or on some alien planet far away? The QRST Ambiguity Principle implies that this critical question cannot currently be resolved.

The Chief Technologist tried to stay calm. It was not every day that one of his rank addressed a meeting of the Chairs of The Corporation. A loud whistling sound signaled it was time to begin the report. He would have to be concise to avoid risking an imperious, mid-report, career-ending dismissal.

"Honored leaders of our planet, exalted and noble, fair and just!" (In a gesture of respect the Chief Technologist looked downward, placed his hands facing up on the cool, silvery table, paused briefly, then looked up and resumed speaking.) "Our species has searched for intelligent life on other planets for hundreds of years, but never developed a clear plan for what to do if we found it. One week ago the search succeeded. Using classical data mining algorithms we detected statistical evidence in electromagnetic transmissions from a sun like ours about 50 light years away. Hypergraphic analysis revealed stunning animated video images of intelligent beings and their artifacts and environment. Like us, these beings have bilateral symmetry, hands with fine motor coordination, and a head containing an apparently large information processing organ. The head also contains sensory organs, whose proximity to the central processor enables rapid reaction times to key sensory stimuli, just like for us and creatures here on our world. There are other surprising similarities as well, such as an audible and written language which we have partially decoded. There are also equally surprising differences."

The Chief Technologist touched 2nd and 4th fingertips to the table, slid them sideways, and the screenwall came to life with a luminous glow. It proceeded to show several short video clips of the alien creatures, as portrayed for their own populace by their own broadcasting technology. The group watched in raptured silence. The Chief Technologist continued, "We surmise from the signals that they have color vision but most likely the colors shown here are different from the colors shown on their own screens, and even if they were the same, they would see them differently than we do. But much more important is their overall level of technological development. We estimate it to be approximately equivalent to ours. How they would react to contact is a key issue since gravitational wormhole travel puts them essentially right next door to us.[1] Travel appears almost inevitable."

The Third Chair of The Corporation looked up toward the dark, richly colored ceiling, squinting as he thought. Who are these beings, with their odd hands so like, yet so unlike our own? With their eyes so like and unlike ours? Cutting semi-familiar silhouettes, yet as startlingly alien in appearance as in our most imaginative holomovies? These beings who dare as we dare, and with similar success, to bend the laws of nature to their will...

The Second Chair of The Corporation spoke next, voice loud and forceful as befit the 2nd-in-command of the planetary ruling body: "Where there is newness there is challenge. With challenge comes danger. With danger, opportunity. The danger is real. Our own history shows numerous instances of new contacts between our civilization, such as it was hundreds of years ago, and the civilizations and societies of others of our species on different continents and islands. Ours became practiced and expert at exploiting, displacing, and even exterminating those cultures and societies. Economic incentives made those who were best at control and exploitation rich, as they swept aside pathetic whiners bleating useless ethical objections. They neutralized their victims and deftly dodged voices of conscience in our own society. Obviously, this alien species may treat new contacts as ours has historically treated new contacts. While contact with them would be different from our past experiences, it would also be the same in important ways. Preeminent among them are the opportunity to acquire great wealth, and the danger of losing all in the risky 'game' of contact."

"Contact with this alien civilization is now virtually inevitable. This is a challenge, it is a challenge we — and probably they — have met before, destroying, prevailing, and of course profiting. There is danger *to* us, but this alien civilization faces the same danger *from* us. There is great opportunity in winning the inevitable clash and profiting handsomely from it..."

A loud BANG announced the First Chair. "If what you say is true, then we have a 50% chance of winning and the same chance of losing. If we win we benefit merely economically, but if we lose it is likely an existential loss. The utility of contact seems well below zero." The Second Chair had anticipated that logic and responded smoothly, with the chill sneer of command.[2] "There is a way. It is best to be the winning side, but almost as good to be the leaders of the losing side if, with great cunning, they sacrifice their side in order to win personally. Military and political leaders, such as ourselves, have always sought to feed the cannons with their minions, proclaiming grand principles and slogans of sacrifice while protecting and enriching themselves and their kin. Why is this situation any different?"

Shocked silence reigned until the Third Chair, returning from a few moments of reverie, responded slowly and deliberately. "Perhaps the spirit of fairness and peace can prevail, yielding a positive expected utility for both us and them. Perhaps we need not be prisoners of this dilemma. Perhaps... it is possible...." There was a pause while the others digested this unfamiliar concept.

The Second Chair broke the silence and continued, confident and matter-of-fact. "It may be possible in theory. But our history shows few such cases. We must have the mindset to win, either as a species, or at least just us as its leaders. If the rest are crushed, so be it."

According to The Protocol of Threes it was now the First Chair's option to speak and render a decision, and all turned their heads expectantly to his raised and glinting station. First's face was an inscrutable mask and several seconds elapsed. Visage still unreadable, First's finger slowly squeezed the control trigger. The gunshot crack of adjournment ended the meeting.

Chapter Nineteen

"Darwin, Meet God." "Pleased to Meet You."

Perhaps spiritual belief will evolve gradually as rational thought and philosophy continue to progress. What form would such progress take? Many people would like religion and science to be reconciled, and indeed that appears possible. In fact the "omphalos" hypothesis solved that problem many years ago, making it, in principle, a non-issue. So why do so few people seem to like this solution? It's not unlikely that, until now, you have never even heard of it. Yet recognition of this solution may grow in the future. In fact the omphalos hypothesis dovetails with another and complementary concept that is gaining in popularity, the simulated universe hypothesis.

Lightning struck at the Creation Museum in Petersburg, Kentucky on Aug. 21, 2013, knocking an employee to the ground. Transported to a nearby hospital, he was fortunately not seriously injured. Some people probably would have preferred lightning to strike an evolutionary biologist instead. Indeed, conflict between religious opposition to evolution and the scientific evidence for it periodically boils over. A case in point is the famous 1925 trial, State of Tennessee *v.* John Thomas Scopes — the "Scopes Monkey Trial." More recent examples focusing on textbooks seem to surface periodically, for example in Texas, which buys enough textbooks to influence textbook content nationwide.

Given that background of social conflict, most people are surprised to learn that, actually, science is readily reconciled with religious belief, even fundamentalist views of creation hostile to evolution. The resolution is achieved by the omphalos (Greek for "belly button") hypothesis, named such because even in a literal version of biblical creation that considers Adam a real person created from scratch, he might well have had a belly button — apparent evidence of a missing prenatal past. The name is from Philip Gosse's 1857 book Omphalos: An Attempt to Untie the Geological Knot.[1] Fifty years later his son commented, "Never was a book cast upon the waters with greater anticipations of success than was this curious, this obstinate, this fanatical volume …. This 'Omphalos' of his, he thought, was to … fling geology [and in particular its fossil evidence of evolution] into the arms of [S]cripture …."[2]

A hundred years after *Omphalos* appeared, Martin Gardner wrote in his 1957 book Fads and Fallacies in the Name of Science, "…it presented a theory so logically perfect … that no amount of scientific evidence will ever be able to refute it."[3] Even the influential evolutionary biologist Stephen Jay Gould weighed in, asking in 1987, "But what is so desperately wrong about omphalos?"[4] It remains under-appreciated and usually ignored to this day by scientists and theologians alike. Most have never even heard of the omphalos hypothesis. One can only ask, "Why?!" Surely the answer is not its irrefutable logic and simple elegance. Maybe it is because it is unsatisfying to anyone focused, not on logic, but rather on one side winning against the other. (What do *you* think?)

Let us learn more about omphalos.

What exactly is the omphalos hypothesis?

Let's start with the belly button question since, as mentioned, *omphalos* is Greek for navel. Some believe Adam and Eve were the first people, created directly by God. Being human, they likely had belly buttons, but not having grown inside a mother, never needed them. If they had them, it would have been evidence of a nonexistent past. Even if someone claimed they did not have belly buttons, the idea of something being created in a way that looks like it experienced a pre-creation past pops up in many other examples. I'm not just referring to new blue jeans manufactured with holes in them — though that does illustrate the concept. Consider the following examples. Were the first trees created complete with annual growth rings? Were the first rocks created with surface wear already in evidence? Did the first oceans already contain dissolved salt and other minerals? Some die-hards might perhaps say no. I wasn't there so I can't refute that (of course, they weren't either). But let's cut to the chase scene, the classic creationist dilemma of buried fossils.

There is no scientific experiment that will ever distinguish between the scientifically motivated hypothesis that fossils are from ancient living things deposited in sediment millions of years ago, and the religiously motivated hypothesis that the sedimentary rock together with its contained fossils were divinely created several thousand years ago (or 100 years ago, or even yesterday along with you and your memories). That is the essence of the omphalos hypothesis.

Other challenging observations abound as well. Human genomes show evidence of different population splits at numerous time points in the ancient past. Different species show genomic evidence of common ancestry at different times in the even more ancient past. Rocks appear to have solidified long ago from hot magma from the Earth's interior in some cases, in others built from layers of sediment deposited over eons, or crystallized in veins out of ancient superheated water in which they were dissolved, etc. Why is the interior of the Earth so hot anyway if not from ancient processes of accretion and radioactive decay? Ancient light from far-away stars appears in the sky nightly. All of these examples suggest an ancient past at odds with literal interpretation of biblical creation. This leads to the dilemma of apparent conflict between science and some religious faithful.

Solving the dilemma

The omphalos principle steps in and with a few keystrokes resolves the apparent contradiction between all these examples of ancient change and the desire of some to believe in fast, relatively recent creation. As Chateaubriand put it in 1802 (over 50 years before Gosse's book), "God might have created ... the world with all the marks of antiquity and completeness which it now exhibits."[5] The universe could have begun 13.82 billion years ago, as cosmologists have concluded; or as some believe, it could have been created *de novo* by God, but in a form that meets scientific criteria for *looking like* it started in a big bang 13.82 billion years ago. There is no way, scientifically, to tell the difference. There never will be. Logically, there *can't* be, because no experiment could ever distinguish between them. That's the omphalos hypothesis in a nutshell, and it really does reconcile the conflict between the science and religion of creation.

Of course, there are some details that could be elaborated. Also various objections have been made and rebutted. For example, it has been claimed that God would never do something like that (the rebuttal: second-guessing God is inherently problematic, because how can anyone really know what God would or would not do?) A web search on "omphalos" will reveal further reading if you are interested, because although the omphalos hypothesis is obscure enough that you likely never heard of it before, it is not that obscure.

With the logical case for the omphalos hypothesis essentially unbreakable, why not call upon it when social controversy pits science against religion? From the standpoint of even a literalist religious creation doctrine, invoking the omphalos hypothesis renders scientific results no longer a threat. From a scientific perspective, creationism is no longer provably a mistake. Yet foolproof though the hypothesis is, both the religious and scientific communities are oddly unenthusiastic about it, as they have they been since Gosse's book was first ignored over 150 years ago. Nevertheless, certain social controversies could become significantly easier to solve.

A mystery — and a prediction

The omphalos principle does what it is supposed to do. It eliminates the conflict between, on the one hand, evolution and other scientific results showing ancient existence and change, and on the other hand, religious dogmas about creation that claim to conflict with science. The biggest remaining mystery, then, is why it is so obscure and little used. Largely ignored in Gosse's time, it remains largely ignored in ours. The various criticisms offered of it are excuses, not real objections, and fail to explain the reluctance to use the omphalos hypothesis to reduce social discord.

The omphalos hypothesis is a clear and elegant resolution to the evolution vs. creation conflict, so it is puzzling that it has been so widely overlooked. Perhaps there is greater interest in maintaining the conflict than resolving it. What is your opinion on that?

Here is a hopeful prediction: Science, philosophy, and religion will gradually grow more sophisticated and wise over time. Assuming this happens we can expect awareness of the logic and social usefulness of the omphalos hypothesis to increase in both the scientific and religious communities, finally leading society to a happy understanding of their intrinsic compatibility.

Religion and the future — Act II

We've just used careful logic to save the scientific and literalist accounts of creation from each other (less literal interpretations of the bible are not an issue). The omphalos hypothesis implies they simply don't conflict. Next, let's save the God concept itself from the usual skeptical argument. It can be done. The result certainly has a non-traditional twist. Yet if a trend toward careful logic characterizes future thinking about religious concepts, many faiths may find themselves increasingly understanding a common, unifying theme. This could happen, or it might be a dream....

Dream worlds

Are you really reading this, or are you just dreaming that you're reading it? How can you tell for sure? You can't, as noted for example by famous philosophers Zhuang (369–286 BCE) and, much later, Descartes[6] (1596–1650). As Zhuang put it, "Once upon a time, I, Zhuang Zhu, dreamed I was a butterfly, fluttering hither and thither, to all intents and purposes a butterfly [...]. Soon I woke, and there I was, veritably myself again. Now I do not know whether I was then a man dreaming I was a butterfly, or whether I am now a butterfly dreaming I am a man."[7] You've woken from dreams before, and though what you are experiencing now likely seems more vivid and realistic, maybe it is just a more vivid, realistic-seeming, and longer dream from which you will at some point awaken into some other reality. Logically, it's a genuine possibility.[8] Personally, I suspect it won't happen. But I could be wrong.

A variation of the dream world idea is that the universe is not your dream but that of another — God. This has intriguing similarities to...

Matrix worlds

In the iconic movie *The Matrix*, ace computer hacker Neo is offered a choice. He could stay within the comfortable world he has always known, but he is informed that it is merely a simulated world (called the "Matrix") inside a powerful computer. Or he could choose to exit the Matrix and experience a bleak, but at least true, reality of rebellion against the intelligent machines who nefariously keep humans' bodies in tanks of liquid with cables connecting their brains to the Matrix. Although *The Matrix* is better known, the simulated reality idea was explored with more care in another movie, *The Thirteenth Floor*, based on Daniel Galouye's 1964 sci-fi novel *Simulacron-3*. That movie envisioned conscious minds made of software and residing inside computerized artificial "worlds."

Today, some Singularitarians seriously want to upload their minds into computers as a way to mentally survive the deaths of their bodies.[9] But when an uploaded mind running on a powerful computer faces the possibility of the computer being turned off, it is facing a genuine extinction of beingness — a real death.[10] Even though a world might be simulated, it feels real to its inhabitants and, indeed, doesn't that make it effectively an alternate reality? This brings us to...

The Sims

This famous computer game runs on a home computer. It simulates a world, populated by artificial people who live their lives under the watchful guidance of the person running the software. The Sims is a rather hollow facade, vastly simpler than what anyone could consider a genuine simulated world. The people in it, for example, are far too uncomplicated to have anything real minds, whether simulated, uploaded, or in any other form. The Sims, as an example of a current, commercial world simulation technology, is just a start. Like other technologies, world simulators will doubtless improve with time. If mind uploading becomes possible, then a future generation of The Sims could contain — Matrix-like — conscious and feeling minds. Such a computer program would be a kind of simulated universe.

Simulated universes

Here are three questions. (1) Could we be living in an advanced Sims-like universe and not even know it? (2) If so, what is the likelihood of such a scenario? And, (3) how does this loop back to the issue of logical thought and its future impacts on religious belief? Let's tackle those questions one at a time.

Might we live in a simulated universe?[11] Consider the evidence so far.

- A dream is an environment simulated by a mind, and philosophers have known for thousands of years that, whether it seems strange or crazy or whatever, there is no known proof that you don't live in a dream. Perhaps you will wake up and find out in a few minutes. Or when you die. Or maybe not. We just don't have any consensus.

- Environments can be simulated not only in a mind (dreams), but also in a computer (*The Sims*, *The Matrix*, *The Thirteenth Floor*). Computers are continually getting more powerful and able to simulate more and more complex environments. Simulations of a complexity akin to what is depicted in *The Matrix* could become possible.

From these points, the most obvious conclusion is that, yes, we might live in a simulated universe and not realize it.

Given the possibility, what is the likelihood we actually do live in a simulation?

Literally billions of people dream every night. That's a lot of simulated worlds. The Sims software can be run by anyone on their personal computing device. If this game gets powerful enough to rate as a genuine simulated universe, then there will be large numbers of simulated universes, running in people's computers instead of in dreams. So, our familiar universe already contains many, many smaller, simulated universes (dreams), and in the future could contain many more (running on computers). And that's just here on Earth — there are billions of extraterrestrial planets out there as well, perhaps containing intelligent races with their own dreams and computers. Consequently, the number of simulated universes (dreams now, and in the future presumably computer simulations as well) is much greater than the number of universes we normally consider real (just one). Thus, if you picked a universe randomly, it would probably be a simulated one, since the collection of universes to choose from has so many more of them.

In the next step the logic gets slightly dicier. You might or might not reach the same conclusion, so think it over and decide. Let us consider our own universe as a random sample drawn from a set of universes consisting of a base universe and the numerous simulated universes it contains. Is ours the base universe or one of the simulations? Very probably one of the simulations, because there are so many more of them. We conclude that our universe is most likely one of the many simulations (dreams, computer programs, or whatever) contained in an underlying base universe. Just as our universe contains dreams and computer programs, our universe would be a dream or program contained in another.

What are the future impacts on religious thought? This question really deserves its own concluding section...

Religious thought and future impacts

A tremendous amount could be said about religion and the future, but here is a *very* concise summary based on this discussion.

- The universe need not be lawful, in that it does not have to follow its own internal physical laws 100% of the time. Miracles and other arbitrary events could occur because *a dream or software simulation need not be 100% consistent.*
- Souls, life after death, even reincarnation[12], parapsychology and other ethereal constructions cannot be ruled out, since in a simulated universe all these details are just parts of the design of the dream or simulation.

- A supreme creator (for example, God) could be defined as the entity doing the dreaming or computer simulation.
- A typical argument against the existence of God is "Occam's Razor." This is the claim that, all else being equal, a simple explanation is more likely than a more complex one. On this view, a complex universe that just exists is simpler than a complex God that not only just exists but also creates a complex universe. This atheist argument is overruled by the observation that all else is not equal: since the universe probably is a dream or simulation, it probably has a creator — the dreamer or simulation software.
- A simulation could start from a starting point (like a "big bang") or, as with dreams, start from any convenient time point with a reality and history already set up and ready to proceed from (as proposed by the omphalos hypothesis). Therefore there is no conflict between science and religion.

For centuries, the omphalos hypothesis has been roundly ignored. Apparently, people prefer to think there is a conflict than to see that there is not. Why? Is it the strange, unreal feel of what is otherwise a logically unbreakable argument, sort of a vague sense of "it sort of *feels* nuts, so it must *be* nuts"? A desire for conflict between religion and science by opinion leaders? Think things over and reach your own conclusions.

Chapter Twenty

In Memory of Daylight Savings Time, R.I.P.

Daylight savings time is a messy patch that no one really likes. People would function better, and so would society, if people lived in better synchrony with the Sun. Normal clocks can't handle such a concept, but computerized timekeeping could. This is what it would be like.

Daylight savings time was proposed in New Zealand by postal worker and entomologist George V. Hudson in 1885.[1] This curious system has since taken hold worldwide, depriving millions of people of a much-needed hour of sleep in the spring. As the saying goes, "Spring forward, fall back."

Alas, issues with daylight savings time abound, while any energy-saving benefits in the modern world are unclear. Readers who deal with young children can see the strain on them when the clocks jump ahead. Adults often feel it too, but generally suck it up rather than be late for work. Clocks also need to be reset twice a year. (If that's too tedious, you could just keep mentally adjusting the stated time by an hour for about 6 months, after which the clock will be right again!) A better solution is smart clocks that adjust automatically, which is increasingly the situation, as on your cellphone. Current smart clock technology is the way to go, but is a mere shadow of what it could be.

It is axiomatic, indeed tautological, that the human organism works best, most efficiently, and with maximum health and energy, under optimal conditions. Like many organisms, we are naturally tuned to a roughly 24-hour cycle. Many studies have shown that working odd hours is associated with health problems, including heart disease, depression, diabetes, and accident proneness.[2] Obviously therefore, scheduling life around an optimal schedule would benefit individuals as well as society. Such an optimal schedule is strongly influenced by the time of sunrise because, as any jet-lagged traveler will attest, the human biological clock strives to synchronize with the natural day-night rhythm of the Earth. The optimal schedule is therefore determined by daylight, not by clocks with or without twice-yearly daylight savings time adjustments.

Clocks can't keep track of daylight because the precise time period of daylight varies continuously throughout the year. This makes it unnatural to change the clock abruptly twice a year to account for daylight savings time, but also makes the non-daylight savings time alternative (used in most of Arizona) of rigidly adhering to a consistent 24-hour clock cycle all year long, also unnatural. Thus the problem goes deeper than just daylight savings time. The real problem is the iron-fisted rule of the almighty clock as it has always existed. Even a properly constructed sundial, its angled gnomon pointing in the exact direction of the Earth's rotational axis, tells the same time as a regular clock, making it astronomically correct but biologically stressful, since the time at which the Sun rises then varies over the course of the year.

"Something should be done about it!" as the popular saying goes. Here's what. The time we get up in the morning, the time work starts, school starts, etc., should be pegged to sunrise. This typically changes by a minute or two (in terms of current time-keeping) from one day to the next. Of course, it would be crazy to have to consult multiple detailed calendars to know whether the office opens tomorrow morning at 7:17 a.m. or 7:19 a.m., whether the kid's school bell rings at 8:03 a.m. or 8:06 a.m., etc. The natural solution is for the bell to ring at (say) 8:00 a.m. every day, just like now, but for 8:00 a.m. to be a constant amount of time after sunrise at the school's location all year long, unlike our current, anachronistically rigid clocks. The current daylight savings time rules could be abrogated and replaced with a simpler rule: When the Sun rises, it is 6:00 a.m.

This strategy, sensible as it is, does lead to a few wrinkles that need to be ironed out. One is that when it is 6:00 a.m. at your home, it won't be exactly 6:00 a.m. at your workplace or the local grade school. Every location on Earth will be its own time zone! But so what? You will learn what time you need to leave in the morning just as easily as you do now, even if your workplace time is a minute or so later or earlier than your home time. No big deal. The hard part is building clocks that can actually figure out when sunrise is and thus what time they should show using this new system. But this is the computer age. Roosters can crow based more or less on when sunrise is and they're not that smart, so surely we can build clocks that know when sunrise is. Clocks merely need to know where they are, whether by using a built-in GPS receiver, by you typing in the latitude and longitude obtained from Google maps or wherever, along with software to compute the proper time for the clock to show based on the location and date. Appointments for telephone conversations or conference calls among widely separated individuals can likewise be coordinated by automatic conversions among participants' times mediated by currently crude, but soon more sophisticated event scheduling software like doodle.com and Google calendar. Long distance travelers would have itineraries too complicated for humans to reliably determine to the minute, but computers simply need to step up to the plate and take care of that. We're certainly not talking about anything beyond the ability of current software technology.

A few more details

The moment of sunrise typically changes from day to day. For example, in Chicago (and many other places) in late March, sunrise advances about 2 minutes per day. That means in our new time system, let us call it "sunrise time," if 6:00 a.m. is defined as the moment the Sun's first ray peaks over the horizon then on, say, March 18, there will be no 5:59 a.m. This is because sunrise on that day is 23 hrs. 58 min. after sunrise on the previous day. Clocks would need to skip 5:59 a.m. and go directly to 6:00 a.m. to greet the sunrise, though hopefully you'll still be asleep and won't care.

The reverse situation occurs in the autumn when the days are getting shorter and, thus, sunrise gets later each day. For example in New York City in September, sunrise is delayed by a bit over 2 minutes each day. That means one minute after 5:59 a.m. the clock can't advance to 6:00 a.m. because the Sun still has not yet risen. The most obvious solution: 5:59 a.m. turns to 5:60 a.m., then 5:61 a.m., and finally 5:62 a.m. Then sunrise occurs and at long last it becomes 6:00 a.m. As a precedent, the 61-second minute is *already* an official part of modern time-keeping: Every year or two a leap second is added just before midnight UTC (Coordinated Universal Time), leading to a minute with 61 seconds (see Figure 4).

Recommendation

Sunrise time should be instituted worldwide. That might take a long time. But there is really no time to waste. As soon as possible would be just in time, because the benefits include a happier, healthier, more productive — and less tired — society.

Figure 4. The US government public clock service, time.gov, showing an unusual time. 6:59 P.M. was 61 seconds long. Its first second was 6:59:00, its 2nd second was 6:59:01, its sixtieth second was 6:59:59, and its last and sixty-first second (shown) was 6:59:60. One second later it turned 7 o'clock.

Chapter Twenty-One

Science and Destiny

Pseudosciences have exerted surprising influence on popular beliefs in the past, but real science is what enables technological advances. So how does science actually work? Here is the view from philosophy of science. A better understanding should enable faster and more efficient technological advancements. That would benefit the economy as well as help make life exciting.

Advanced technology would be literally unthinkable without the power of science to explain how the world works. Yet such power can be dangerous — as in the power of the atom, used in nuclear technologies — because, in the well-known words of Briton Lord Acton (1834–1902), "power tends to corrupt." So it is useful to understand what science is and is not, as well as how it has been used and misused in the several hundred years since the modern conception of science began. Those lessons will likely continue to apply for as long as science exists. Hopefully that will be for a long, long time. These understandings carry the potential for guarding against the dangers of pseudoscience, as well as the misuse of science for ill.

To achieve these understandings, we turn to philosophy. Ultimately though, understanding alone is not enough, and actions that affect society must be taken. Deciding on no action is an action, and indecisiveness leading to no action is also an action. Action is the domain of those with the power to act, including government, business, and, in a democracy, the voting public. Philosophy of science can illuminate and inform, thereby enabling and (hopefully) encouraging wiser actions.

We will discuss what science is, then pseudoscience, and then how science can be done better. If good science can help society make good decisions, then better science should enable better decisions. Since the future of technology and its most dramatic long-term possibilities depend on the future of science, better science ultimately means more and sooner startling technological successes. Some Singularitarians suggest that the ultimate potential of technology might be revealed in less than a hundred years, though fully achieving that revealed potential might take a bit longer. Perhaps the limits of what will be achieved will be reached within a thousand years. But that is just a guess. Of course, if the human race goes off course it may never happen, at least not here on Earth.

An entire book could be written on what science is and many have been. We summarize in just a few paragraphs next (the reader may if desired skip directly to the section "**Pseudoscience is not science**").

Science and solipsism

What makes our world and our universe work the way they do? Science tries to answer that call to seek understanding of the workings of the physical world. But this assumes that there really is a physical universe. Do you think the universe exists? A few people — solipsists — have argued that maybe it doesn't.[1] Solipsism holds that the universe could be merely in the mind and imagination of the observer, and there is only one observer, you (or, from the standpoint of a solipsist, the solipsist).[2] For example, maybe you are just dreaming your life and the universe.

Though odd, it is not immediately obvious why (or even if) solipsism is wrong. One approach is to notice how much you do not know but can find out. It seems more sensible to think you are actually learning something you did not know, rather than just dreaming it up. Learning something in school that is hard to wrap your mind around is an example. If on the other hand you do just imagine everything (even this paragraph), then your mind is a lot smarter and more imaginative than you normally give it credit for! Perhaps the main implication then, if solipsism held, would be that because your mind is inside your skull and the universe is inside your mind, the universe would end upon your death. But there are people who have argued exactly that and are no longer alive, and the universe still exists. More fundamentally, if the universe was inside your mind, then your mind would have to exist in a larger universe containing it. So let us agree, as a start, that the larger universe exists.

But the existence of the universe is not enough for science. Science also needs the universe to work in a consistent way, because the scientific process is aimed at discovering the orderliness of the universe. The mind is not so constrained to consistency, but seems to work better when it does a somewhat credible job of modeling the orderly universe. That can explain why, for example, the brain's default network, the connected set of brain regions active during undirected wakeful states, normally prevents one from hallucinating unreal things. When this network is suppressed, as by psilocybin, the hallucinogenic ingredient of "magic mushrooms," strange things are vividly perceived, but they don't really exist.[3]

Luckily the present state of science has shown considerable success in understanding things that do exist, so it is clear that the universe is orderly enough for science to work and that science is capable of discovering how it works a good deal of the time. Science does this empirically. That means by observing things, such as the results of experiments. This contrasts with, for example, discovery by seeking mystical enlightenment.

Science and induction[4]

Science is fundamentally about understanding reality by observing it. (This is called "empiricism."[5]) More technically, discovering the structure and workings of the universe requires generalizing from past observations to make predictions about future observations. This type of reasoning is called induction. It is different from deduction — reasoning forward from first principles — which is what mathematics emphasizes and why mathematics is not a science. Thus, observing several dozen things falling down after being tossed upward, you might induce the theory that what goes up must come down. Observing a fly ball in a baseball game eventually descending, you take this as confirming evidence for the theory. Yet, a helium balloon tossed upward may not be observed to come down, and a powerful enough rocket really will never come down. Thus conclusions reached by induction are not infallible because there is always the possibility that a disconfirming observation might come along later.

Karl Popper (1902–1994), a philosopher who is highly regarded among scientists, addressed the fallibility of induction. His solution was to characterize science as progressing, not by confirming theories, but by falsifying them. Thus upon observing a helium balloon that does not fall, the theory that "what goes up must come down" might be modified to explain why things lighter than air behave oppositely. Observing fast rockets that escape into space might then lead to the concept of escape velocity, and observing the Moon might lead to the concept of orbits (in which falling toward the ground happens at the same rate that the ground of the round Earth curves away from the falling object, so it never lands). Observing other planets, stars, and moons orbiting each other falsifies the theory that the Earth sucks and leads to the concept of Newtonian gravity. Further observing gravitational bending of light falsifies Newtonian gravity, resulting in Einsteinian gravity, which is basically where we are now. On this view, the search for scientific truth is deeply dependent on falsification as a strategy.

If science involves trying to falsify theories rather than confirm them, then concepts that cannot be falsified by observation are not part of science. For example, there is no observation that could ever falsify the claim that God exists, so God's existence is simply not a scientific question. (On the other hand, the theory that God prevents injustice can be addressed scientifically because one can look for counterexamples that would falsify such a theory.)

Unfortunately for Popper, the point of science is not figuring out how things don't work, but rather how they do. This requires addressing how new theories come to exist.

Thomas Kuhn

Kuhn (1922–1996) found that the scientific search for truth is a community effort. The people in a scientific field may commit to a major theory as a working assumption and work on testing and detailing it. Relativity in physics and plate tectonics in geology are examples. If and when falsifying observations build up to the point where they cannot be dismissed, someone may then propose a new major theory and a "scientific revolution" occurs. Hence the title of Kuhn's influential book *"The Structure of Scientific Revolutions."* Examples of scientific revolutions include relativity displacing Newtonian mechanics, and plate tectonics (in which continents move) displacing the contracting Earth theory. When such a revolution occurs, younger scientists tend to commit to the new theory while the old dogs, mostly unwilling or unable to learn new tricks, gradually fade away like old soldiers.

Kuhn's emphasis on the conduct of successful science as a social enterprise occasionally contrasts with the imaginations of science fiction storytellers, as when they depict secretive, misanthropic mad scientists as lone rangers with mad and diabolical plans. It also de-emphasizes the concept of falsification as the road to truth. This is fortunate because of the intuition that there must surely be a role in scientific inquiry for confirmation rather than only falsification.

After Kuhn

Imre Lakatos (1922–1974) and Larry Laudan (1941–) built on Kuhn's concept of science as a social community, noting for example that some scientific fields contain more than one theoretical paradigm operating at once. Psychology, for example, has a history of multiple concurrent competing schools of thought. Paul Feyerabend (1924–1994) argued for no scientific methodology at all in his book *Against Method*, stating at one point "… the principle: *anything goes*."[6] Luckily that extreme position seems mostly for rhetorical effect, since he clarified that with the much more reasonable "… *all methodologies … have their limits.*"[7]

Scientific progress can also hinge on the demographic characteristics of the scientific community (as well as the interests of individual scientists). For example, understanding of the complex roles of females in various primate species has been attributed to the presence of women in the primatology field. These social characteristics of the scientific community are good news for prospective scientists attracted to working on problems that combine imaginative thinking about old and new paradigms with the more analytical and detail-oriented activities that the general public already associates with science.

Even Popper's falsification methodology needs some updating, because science can address confirmation directly by a technique called Bayesian statistics. Given an existing ("prior") probability of a hypothesis a new probability can be derived from new evidence. The problem is that there is frequently no way to know the prior probability. Without that starting point the ending point tends to remain somewhat mysterious. Yet confirming evidence is confirming evidence, even if it is not clear precisely how confirmed the underlying hypothesis has become.

Science vs. pseudoscience[8]

So what then is science? If the above presents science on the half shell, then here is science in a nutshell: It is an approach to investigating questions of some importance that are answerable by experiments and other verifiable observations. Centuries of science have developed and refined this approach into something that, by and large, works. It works because newer theories that have been tested and accepted usually work better than those they replace. We know they are better because they explain experimental results better and, importantly, because they lead the way to the technologies we now possess.

Now that we have the "What is science?" concept under control, it becomes easier to see pseudoscience for what the name suggests it is: non-science whose proponents say it is science. Pseudoscience has many and varied examples, all rooted in the assumption by the proponents that science provides truth. As we've seen, that assumption is not quite accurate, since science actually provides theories along with varying amounts of corroborating and disconfirming evidence for them.

Creationism, and its variants like creation science and intelligent design, are a perhaps particularly unfortunate example of a pseudoscience. I say unfortunate because creationists could more logically claim truth using philosophy as a tool than science (as explained in chapter "Darwin, Meet God." "Pleased to Meet You."). So there is no conflict between religion and science, and hence no reason to fight over the science of evolution.

Climate change denialism is another example of pseudoscience because its community lacks critical peer review, is mostly divorced from the technical training that is part of the climatology field, and ignores experiments in data analysis that refute its central hypothesis.

Then there are people who believe fluoridation of drinking water is a secret government plot. They are generally not members of the public health or dentistry communities. Also, some of them have bad teeth. Moving into the farther reaches of nuttiness, Royce and Zolot claim that Satan created the dinosaurs. Pseudoscience. Arggh!

Immanuel Velikovsky, originally a successful medical researcher, became an independent pseudo-scholar of astrophysics by spending years camped out daily in the Columbia University library. He might never have created even a ripple in space-time if not for his uncanny ability to write catchy titles for his books (like *Worlds in Collision, Ages in Chaos, Earth in Upheaval, Mankind in Amnesia, Cosmos without Gravitation,* and *Stargazers and Gravediggers*). Again, arggh!

Pseudoscience is harmful because it fails to produce better understanding or new and useful technologies; rather, it impedes real science and hence impedes the advance of both understanding and technology. It also tends to make society work less well by lending an element of irrationality to decision making. Aaarggh!

Recommendations

If scientific progress proceeds more or less as it has been doing, it is not a tragedy. But it could advance faster and better with a few strategic changes. One way to speed progress is to have more scientists. It is society's decision whether that happens or not. It is expensive but may ultimately pay for itself — or not. Research (scientific, of course) sufficient to answer this question should be commissioned by society.

Another way is to improve the productivity of scientists. Who could argue with a higher rate of scientific progress per scientist? Here are some strategies to help make that happen.

- All student scientists are taught at least something about scientific methodology, but its philosophical underpinnings are generally taught desultorily or not at all. Teach philosophy of science to students as part of their education in scientific methodology. This will help make their work more significant, and significance is a crucial measure of scientific value.

- Scientific papers and research proposals usually describe the motivation; in other words, why the work is needed. The motivation is closely related to the significance issue just discussed. Those motivating passages can refer to philosophical foundations (for example, is a paradigm or paradigm shift at issue?). This will make motivations clearer and thus higher in quality. Student scientists who are taught these philosophical foundations will tend to write this way themselves and expect it of others.

- Scientists are produced by an apprenticeship process performed under an advisor. Yet scientists are rarely taught how to be good advisors, and there is too little accountability for poor advising, resulting in dysfunction in the creation of the next generation of quality scientists. One student's brilliant and profitable insight was stolen by his advisor. He left the program and spent a long time sitting on mountain tops contemplating life — and society has one less creative chemist. Another advisor demanded his graduate students staff the lab with hours that drove his students to the brink of throwing in the towel — even when he was on the vacations he disallowed his students. A third advisor demanded students to publish if they wanted to progress — and provided little or no assistance in doing so even though the students did not know how. The occasional advisor keeps student scientists from graduating because of the personal career value of keeping highly skilled but surprisingly low-paid graduate students stuck in the lab (the only half-joking term "graduate student slave" is well known in the academic community). PhD and other doctoral programs and even individual advisors and laboratories should have the opportunity for, desire of, and expectation of accreditation, ISO (International Standards Organization) certification, or some other credible documentation that they meet quality standards of practice in producing new PhDs. Also, the expression "graduate student slave" should be banished from the scientific humorous vernacular. More importantly, the ways in which it is occasionally practiced should also be ended.

- Scientific progress is mostly vetted and recorded in peer-reviewed papers. There exists a continually expanding body of such publications, but there would be even more if this body started including items that can be important in advancing the field but are currently under-disseminated — unpublished, published but without effective peer review, or discouraged from being written at all.[9] Such items include negative results; position papers; creative, stimulating, and imaginative speculations; and updated (not necessarily by the original authors) versions of previously published papers to keep them relevant. The modern online publication paradigm makes such items easier than ever to disseminate: Web pages are cheap to publish and easy to access. The result would be more publications that advance knowledge for others to build on further — hence more progress.

The faster science works, the faster we'll get where we're going, wherever that is. This should be empowering, but may actually lead to faster, more impulsive deployment of risky, self-destructive new technologies. Consequently the science and engineering of rational behavior by human societies is particularly in need of expansion.

Do you want to explore the mysteries of the universe... to push forward the bounds of human knowledge... "to boldly go where none have gone before"[10]... to wrest the most satisfying and nutritious secrets from the jaws of nature? Then you might find becoming a scientist a bit anticlimactic. Most scientists spend their professional lives not revolutionizing human knowledge or even building the next paradigm shift but, metaphorically speaking, scratching in the dirt for a few nicely colored pebbles. On the other hand, you never know what could happen. So it is with science as with so many other human endeavors.

The Third Generation:

The Next Ten Thousand Years

Chapter Twenty-Two

The Teeming Cities of Mars

The law of exponential population increase means that even a small Martian colony could expand rapidly, overpopulating the entire planet with billions of people in just a thousand or so years.

Would you accept if offered a once-in-a-lifetime opportunity — to be in a small group starting a new offshoot of humanity, the first permanent, self-sustaining colony on *Mars*?[1] Mars, the red planet, captor of the imagination and stuff of ancient myth![2] Most people don't get such a rare chance, so this may sound like good news indeed. If you're ready to sign up, Mars One started taking applications in 2013 for astronauts for a 2023 colonization mission (Figure 5)[3]. But first consider the bad news. For one thing, a 2014 analysis showed that their life support technology is not yet reliable.[4] Hmm. And that's not all.

Figure 5. Proposed Mars colony (credit: Mars One / Bryan Versteeg).

Travel technology will be able to manage a one-way trip to Mars before it can swing a round trip. Since you're on the first mission, once you leave Earth you cannot change your mind. There is no coming back. Realistically, once there you're there for a long, long time, probably the rest of your life, like it or not. That is because it is hard to get something all the way to Mars, and the spaceship will have to carry not only the colonists themselves, but also everything they need to build a self-sustaining, permanent habitat. It would be technically challenging and expensive to carry along the fuel, extra hardware, and so on needed for a trip back to Earth, and if the ship could carry all that, why not carry more hardware with which to establish the colony instead? Besides, since the purpose is a permanent colony, there's little point in a planning a return trip anyway. Even the ship itself would be repurposed to be an essential part of the colony habitat, since flying it back would not be planned.

Repurposing the ship would only be a starting point, however. Manufacturing equipment would also need to be brought along capable of producing the materials needed for the colony, such as glass or plastic for domes to grow crops, live, and play in.[5] If you're lucky. It might sound pleasant to walk through stands of crops growing plump, delicious foods as you care for them and watch them grow, all while taking in Mars' magnificent, deep-colored sky, and red-tinted hills and plains through the large transparent dome. Not bad at all! However, the reality might be more prosaic. It might be much easier and more efficient to have insulated clear-topped tanks for growing algae, while living in much more cramped quarters instead of domes, perhaps underground without a view. Chemical synthesis of required human nutrients using solar powered devices might work even better than algae tanks and require even less human contact with the surface.

Another piece of bad news you should be informed of is the roughly 2–4% risk of mission failure.[6] That dry statement hides a stark reality: It means everyone dies. If something goes seriously wrong in the colony, it is likely that the folks back on Earth would be unable to help much.

Before making your decision to go or not, keep in mind that even if everything goes as planned there would be little chance to get away from things for a while. Perhaps there will be an occasional space-suited jaunt into the barren, dull, reddish landscape, desolate from horizon to horizon. If you go, it is because the idea is exciting, not because daily life will be better than on Earth, because it probably won't be.

On the other hand, modern hand-held computers loaded with everything from games to Wikipedia could be brought that would satisfy an unlimited thirst for that sort of diversion. Some sort of access to the web is certainly possible as well, via radio communication with Earth. That would render the term "World Wide Web" not only incorrect, but rather parochial. Interplanetary Web, anyone? Imagine growing up in such a tiny, isolated outlier society with all of one's understanding of Earth and everything on it obtained from a computer. This will be the impoverished experience of the next generation colonists, born and bred on Mars — indeed, genuine Martians.

Even that modest window on Earth would need electronic devices, which will eventually break and be difficult to repair, although shipment from Earth of small, lightweight digital components might be a possibility, especially if a trading relationship were established. There would certainly be a market here on Earth for at least a small number of high-priced Martian rocks and the like. It might also be possible to jury-rig access to at least some basics on the web like static text documents using simple homegrown electronics.

In any case, a cool-headed assessment suggests the quality of life on Mars would be a lot lower than life here on Earth for the typical reader. So, would you go?

Informal polling shows that many college-age males would go for it, provided enough females were going too. Females are often more wary. Your mileage may vary, but as Mars One applicant Kellie Gerardi enthused, "... either you get it or you don't ... I don't know if ... I could ever explain that to you."[7]

Teeming cities

A twenty-person colony is not a teeming city, though it may *seem* teeming, what with crowding into living space that is scarce owing to the difficulty and expense of building each new square foot of high tech, hermetically sealed, oxygenated habitat for housing colonists and producing food. Expensive or not, square footage will need to be constructed because, if all goes well, children will be born and the colony will grow.

Natural growth rates for human societies vary, but are generally under 5% per year. Overall, world population is currently growing by about 1% per year. Let's assume for a moment that our Martian colony experiences the same growth rate, 1% per year. How long do you think it would take for the original twenty person colony to expand into a vigorous town of a thousand people? 100 years? 500? 1,000? 5,000? The answer can be readily found with a calculator or spreadsheet: just 394 years. How long for the colony to become 10 million Martians — a teeming city or, more likely, several? Take a moment to guess. Here is the answer, spelled backwards so you don't read it accidentally. .sraey neves ythgie dnasuoht eno erem A

Why stop at 10 million? Population growth on Earth didn't, and there is no reason why it would on Mars either. A burning question then becomes when Mars will pass its capacity and tip into overpopulation. If capacity is 10 billion people, er, Martians, it would pass that point, starting from the original 20 colonists, in only 2,014 years.

Here on Earth, the road to 10 billion has already taken a lot longer than that. How long? The question is unanswerable because we don't know when the process started. But perhaps a rough starting point could be assigned based on what we do know. For example, we know that the surprisingly low genetic diversity of humans (compared to most species) suggests we "began" (in a sense) relatively recently and have not had time to accumulate the mutations needed for much genetic diversity. The Toba supervolcano eruption about 73,000 years ago has been proposed as this starting point, by causing a multi-year volcanic winter from throwing so much dust and smoke into the atmosphere. On this view, populations of protohumans were devastated, leaving only a small community alive. That group then expanded, sweeping across the world. Adding in a modest amount of Neanderthal blood (up to 4% is typical), and factoring in the evolutionary changes since then, we get the human race.

Whether human colonization of Earth began 73,000 years ago or at some other time, it is clearly taking a lot longer to reach a population of 10 billion here on Earth than it would on Mars, given even a modest 1% annual population growth. This is due to the scourge of infectious disease — "pestilence" — as well as other privations. Without those curbs, populations have been often observed to expand at rates in the 3–4% per year range. So our 1% growth rate assumption for Mars may be too low. Consider an optimistic growth rate of 3.5% instead. Now our lively little town of 1,000 happens not in 394 years but in a mere 114 years. 394 years instead gives us not a town of 1,000 as before, but a teeming city of over 15 million inhabitants! A mere 583 years suffices to hit the 10 billion mark. It's not hard to figure this out yourself with a spreadsheet.

What Martians can do

Without uncontrolled infectious diseases brought from Earth to contend with, courtesy of modern civilization, Martians will be in a good position to quickly populate their new world. Food production and other necessary technologies will be somewhat solved problems right from the beginning, or the colony could not even get started. Thus, avoiding Malthusian overpopulation scenario in the end, and population vs. resource growth mismatches along the way, will likely be real issues for Martians who seek to keep their civilization and planet in good shape.

If you, a Martian, are reading this, I advise keeping in mind the often-observed fact about societies on Earth that a high standard of living in conjunction with readily available contraception are critical factors in holding back undesired population growth and resultant overpopulation. Similarly, a high standard of living can be promoted and maintained by controlling population so that plenty of Martian resources are available to everyone. Earth itself is an exemplar of the human population dynamics and management pattern that may be expected to apply on Mars as well. Earth may be merely humankind's first step to colonizing the cosmos; it would be best to make the experience of Earth, as well as Mars and its teeming cities, a good model for colonization of the solar system and the stars beyond.

The Fourth Generation:

The Next Hundred Thousand Years

Chapter Twenty-Three

Big Ice

Global warming, if not controlled, will have major effects for hundreds of years to come. Yet over a much longer term, we're actually in an interlude between glacial deep freezes, and glaciers will most likely descend again.

As World War I broke out in 1914, an uncontroversial Serbian professor at the University of Belgrade suddenly found himself under arrest. Milutin Milankovitch, a studious academic, was deemed a potential danger by the Austro-Hungarian government owing to his suspect nationality, though a less likely person could hardly be imagined. Subsequently transferred to a loose house arrest in Budapest, Hungary, ready access to the Hungarian Academy of Sciences library enabled him to lay the foundations of his life's work: understanding the interaction between Earth's motion in space and the periods in the distant past when glaciers descended from the North Pole — times popularly known as "ice ages."[1]

The broad current of truth he found is still with us today. It constitutes the theory that very long-term cycles in the Earth's motion cause enormous ice sheets, often a mile or two (couple of km) thick, to descend from the north, covering the land and obliterating everything in their path. Their weight pushes the land downward and their motion grinds new valleys out of solid bedrock, leaving new features from the Great Lakes to fertile soils made of pulverized rock. The same cycles also govern the periodic retreat of these glacial ice sheets. Glacial advances and retreats have been occurring for millions of years, and they are projected to continue for millions more until the continents themselves drift away from their present positions. In honor of his work, these periodic motions of the Earth and the shifts in position of its orbit in space are collectively named Milankovitch cycles.

There are three major cyclic periods with particular relevance to glaciation, ranging from 21,700 years long to about 100,000 years. Let's dig a bit into how they work.

The shortest Milankovitch cycle

The Earth spins like a top. At a rate of exactly one revolution per day, during the daytime your spot on the Earth is, whether obliquely or directly, facing the Sun, while at night it faces away. Like a top, the Earth has an axis of rotation. This axis is the foundation (though not the full story) of the shortest Milankovitch cycle, because it *precesses,* or wanders cyclically. This process is known as axial precession or, commonly, just precession (even though there are other kinds of precession too, as we will see).

Axial precession: wandering of the axis of rotation
    Consider a spinning top. As it slows down, the stem, representing its axis of rotation, begins tracing a circle. You probably noticed this as a child. If you have one handy, why not spin it now and see for yourself? This is called axial precession. The time period for this is long enough that the top spins many times during the time it takes the axis to move around its circle once. While a toy top may precess all the way around in a second or less, bigger tops precess more slowly. The Earth is a very, very big top. Its axis is currently tilted away from vertical by 23.44 degrees (where "vertical" is straight up and down through the plane of the Earth's orbit). The Earth's axis precesses clockwise (viewed from above the North Pole looking downward) one time around about every 25,772 years. If the Earth was the only object in the solar system besides the Sun, those 25,772 years would constitute a Milankovitch cycle (see

youtube.com/watch?v=J9Chu4-VlT0). If only it was that simple. Instead, another effect, orbital precession, changes the effective length of the axial precession cycle.

Orbital ("apsidal") precession

There are seven other planets besides Earth, as well as additional smaller non-planetary bodies like Pluto, and their gravities all affect Earth's orbital behavior to some degree. One effect of this is that, besides the aforementioned axial precession, there exists another kind of precession, orbital precession. To visualize orbital precession (usually known among astronomers as apsidal precession), first picture the orbit of the Earth. It is an oval (technically, an ellipse) around the Sun with the Sun nearer to one end than the other. The end nearer the Sun is called the perihelion, 'peri-' meaning "near" and 'helios' meaning sun in Greek. The other end, farther from the Sun, is called the aphelion ('ap-' from the Greek 'apo,' "away"); see Figure 6.

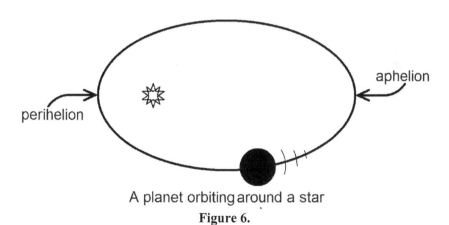

A planet orbiting around a star

Figure 6.

The easiest way to picture orbital precession is to imagine the Earth proceeding along its orbit, moving counterclockwise at one year per round trip, and reaching its aphelion (point of greatest distance from the Sun) at some point during the year. The Earth then continues on its way. Close to a year later it has gone all the way around the Sun and is back at its aphelion again. But the aphelion is no longer in the same place! It has moved a tiny bit counterclockwise because the entire ellipse has rotated counterclockwise. That's why it took very close to a year, but not exactly a year. If the aphelion was at 9 o'clock this year (as in Figure 6), then it would have precessed to 8 o'clock in about 9,330 years. In fact for the aphelion to precess all the way around once requires a good 112,000 years. I find the image of the relatively far away aphelion gradually swinging counterclockwise around the Sun as the oval of the orbit rotates counterclockwise more vivid, but the perihelion, opposite the aphelion and the point on the orbit nearest to the Sun is doing exactly the same thing (see Figure 7).

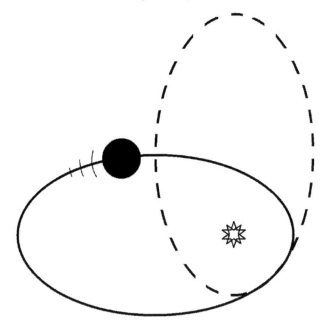

Figure 7. Viewed from above, the orbit has precessed counterclockwise from the old orbit (dashed) to the new (solid).

Orbital precession does not itself create a Milankovitch climate cycle but, acting in concert, orbital and axial precession do. In brief, while the perihelion gradually marches counterclockwise, once around in 112,000 years, the Earth's rotational axis marches clockwise around its own circle once every 25,772 years. About every 21,700 years, the Earth's axis's tilt is toward the Sun at the same time of year that the Earth is also at its perihelion. (By "tilt is toward the Sun" is meant that the Earth is tilted so that the North Pole is not facing perfectly straight up but rather slightly toward the Sun, and the South Pole therefore faces slightly away from the Sun.) Half a year later the Earth is on the other side of its orbit, the Earth is at aphelion, and while the axis is pointing in the same direction as it did before, that direction is now leaning away from the Sun, instead of toward it, because the Sun for practical purposes is on the other side of the Earth. (Actually the Earth is on the other side of the Sun but, relative to each other, this is the same thing.) This 21,700-year periodicity is a Milankovitch cycle, because where we are in that 21,700 year period affects the climate. Here's how.

Suppose for a moment that the Earth's axis of rotation was exactly vertical, at a perfect right angle to the plane of the Earth's orbit and thus not tilted at all. Then the Sun would shine squarely on the equator from directly overhead at mid-day, and if one considered points north or south of the equator the Sun would shine more and more glancingly on the ground as the surface curved backward from the equator. Thus the equator would be hot from the Sun passing directly overhead at mid-day, and the climate would get cooler as one traveled north or south, because the Sun's most overhead position (at mid-day) would become progressively farther away from directly overhead (that is, the sun would stay closer to the horizon).

Next suppose instead that the upper (northern) end of the axis tilts toward the Sun, as is the actual case for Earth for part of the year. Now the Sun shines most directly, not on the equator, but on a point north of the equator (Figure 8).

This tends to heat up the northern hemisphere, so in the north, we call this time period "summer." But south of the equator, it's winter, because with the Sun shining squarely on a point north of the equator, south of the equator the Sun is low in the sky, hence not blazing directly overhead, and therefore cool. On the other hand, a half-year later the axial tilt will be the same, but now leans away from the Sun because the Earth is on the other side of its orbit (Figure 9). We call that "winter" (in the northern hemisphere, but it is summer in the southern hemisphere).

Figure 8. The Sun is shining most directly on the northern hemisphere, so it is summer in the north and winter in the south.

Figure 9. Six months later the Earth is on the other side of the Sun, which now shines most directly on the southern hemisphere, so it is summer in the south and winter in the north.

Nowadays, aphelion occurs during northern hemisphere summer, so summers are on the cool but long side while winters, occurring at perihelion, are relatively warm and short (conversely in the southern hemisphere, winters are colder than they would be otherwise and summers warmer). Northern summers will start warming and shortening, and winters cooling and lengthening, while the opposite will occur in the south. This process is too slow to make a noticeable difference in your lifetime, and is dwarfed by the current global warming trend caused by the greenhouse effect. Although this cycle takes a hefty 21,700 years to repeat, it is merely the shortest Milankovitch climate cycle.

The medium-length Milankovitch cycle

The theory that cool Northern summers tend to trigger the ice age descent of glaciers from the far north across what is now normal land is hard to square with the 21,700 year cycle as a trigger, in part because the timings of glaciations do not seem governed by a 21,700 year pattern. The theory is better supported, however, by a 41,000-year Milankovitch cycle in something called "obliquity." I will explain obliquity. But first, consider its importance.

About 2.6 million years ago the continents had drifted into an "ice age position," putting the Earth officially in an ice age. In this ice age, glaciers advance, covering large amounts of land, then things warm up and they retreat, prior to advancing once again. That ice age still continues today, but luckily we are in a warm spell between glacial advances. The current warm spell began around 11,000 years ago and will end in the future, perhaps in 50,000 years, perhaps farther off than that. That's right, we are in an ice age right now (often the term "ice age" is used colloquially to refer to a single period of glacial advancement, but technically the term for such a period is "glaciation"). Prior to the current ice age there were others at different very ancient times in Earth's prehistory. The worst of them are thought to have caused "snowball Earth" states, where ice coverage extended clear to the equator. But details about ancient ice ages hundreds of millions, or billions of years ago, are tricky to nail down for sure.

When the current ice age began, its glaciations occurred fairly regularly, every 41,000 years! Scientists tried to make sense of this remarkable fact using, naturally, the effects of changes in obliquity, because it has a 41,000 year cycle too. Obliquity is a characteristic of the axial precession described earlier. Recall the stem of the spinning top. As the top slows down, the stem begins describing a circle: axial precession. The circle gets wider as the top slows, until finally it is so wide that the side of the top touches the floor and it promptly crashes on the floor. In the case of the Earth, which doesn't slow down appreciably, this circle gets bigger, but then smaller, then bigger, smaller, on and on, with a period of 41,000 years. When the circle is big, it means the Earth's axis is more tilted than when it is small. The amount of tilt is called the obliquity. At its widest, the axis is 24.5 degrees away from vertical and, at its smallest, just over 21 degrees, though most cycles do not quite reach those extremes. This range of about 3.5 degrees might not seem like much, but to the finely tuned climate system of Earth it means a lot. Here's why.

Cooler northern summers are believed to encourage glaciation by not melting off preceding winters' snow covers. As a consequence the otherwise dark ground may stay white year-round, therefore absorbing less sunlight, compounding the coolness problem and making it even harder to melt the white snow coating the next summer. The result is a "vicious cycle," or positive feedback loop, and things can get a lot worse before they get better — as in creeping sheets of solid ice, a mile (1.6 km) or more thick, covering the land. This has happened countless times in the past, many within the current ice age. Borrowing the words of Norse myth, "The congealed venomous streams continued to send out frost."[2]

What is it about the obliquity cycle that causes summers to cool, warm, and then repeat, every 41,000 years? The reasoning is just a little simpler than for the 21,700 year cycle. Summer is defined as when the axis tilts toward the Sun. In the northern hemisphere, that means the higher latitudes face the Sun more directly than in other seasons. The Sun beats down more strongly because it is more directly overhead. Days are longer — up to 6 months without a sunset at the North Pole. The result is more warmth. That's why we call it summer. This effect is accentuated by greater axial tilt: Greater tilt, warmer summers. Therefore less tilt leads to cooler summers, which melt less snow and ice, allowing glaciers to form, move, and cover the land. Every 41,000 years... until about 800,000 years ago.

The pattern changes: A longer Milankovitch cycle

Eight hundred thousand years ago the alternating glaciations and warm periods changed from a 41,000 year cycle to what looked more like a 100,000 year cycle time. Although the amount of obliquity still cycles at 41,000 years, a third Milankovitch cycle appeared to have taken control. This third cycle is about the amount of eccentricity in the Earth's orbit.

The Earth's orbit is sometimes somewhat eccentric (elliptical rather than circular) and other times it is nearly circular. Currently its eccentricity is 0.0167 and will trend toward 0 (circular) for the next 26,000 years or so. When it is close to circular, the Earth gets the same amount of sunlight overall, year-round. When more eccentric, the Earth and Sun are closer for part of the year but farther away for another part and the Earth gets more or less sunlight accordingly. Currently the Sun is about 7% brighter at its nearest approach during the year compared to its furthest.

A cycle through both the more circular and the more eccentric phases takes about 100,000 years. Furthermore, though not studied by Milankovitch, the tilt of the plane of the Earth's orbit (the "ecliptic") varies as well, a process called precession of the ecliptic or planetary precession, with a period also of about 100,000 years. Ice age glaciations have occurred about 100,000 years apart since 800,000 years ago. Far from solving the ice age timing problem and thus enabling prediction of the next one, however, this leads to puzzling questions.

Question 1: The effect of the 100,000 year cycle on solar energy input at 65 degrees north — something of a pressure point for the northern hemisphere from the standpoint of glacial descents and retreats — is small compared to the effects of axial precession and obliquity. So why would it control the glacial cycle?

Question 2: The 41,000 year obliquity cycle characterized the glaciation cycle until 800,000 years ago. So why would it stop?

Trying to answer these questions has attracted considerable effort in the paleoclimatology community. However there is growing evidence that, in fact, the 41,000 year obliquity cycle actually still dominates. For example consider the glacial era preceding the most recent one ending 11,000 years ago. That earlier glaciation tipped into its own termination phase 123,000 years earlier. 123,000 years is three 41,000-year obliquity cycles.[3] To explain the apparent 100,000 year cycle one need merely posit glacial retreats at intervals of 2 or 3 obliquity cycles (about 82,000 and 123,000 years apart), averaging about 100,000 years and thus by coincidence giving a false impression of being driven by the 100,000 year eccentricity cycle.[4] As for why obliquity is so important, it appears that glacial retreats are triggered by a peak in the total summer solar energy impinging on the upper northern hemisphere (65 degrees in latitude is the typical proxy for that). This peak happens at highest obliquity — every 41,000 years.

Clearly though, obliquity cannot be the full explanation because glacial retreats averaging 100,000 years apart have to skip many obliquity cycles. The whole story is thus more complex. Something — perhaps very gradual cooling over the past few million years (punctuated by the current global warming spike caused by human activity) is causing glacial retreats to skip obliquity cycles (Figure 10)[5] Furthermore, the solar energy effects of all the Milankovitch cycles act simultaneously, sometimes adding together and sometimes tending to cancel out. This leads to considerable variation in the degree of solar forcing of glacial terminations, even at the peaks in the 41,000-year obliquity cycle (Figure 11). Also, keep in mind that vast glacial coverings cannot retreat unless they build up first, and the buildup occurs slowly compared to the more dramatic retreat events.[6] Buildup, too, is influenced by Milankovitch cycles.

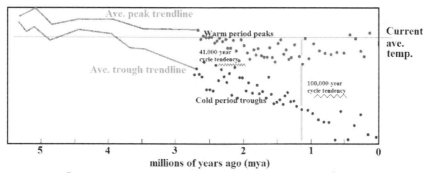

Figure 10.[7] Major global temperature peaks and troughs. These have become less frequent, while increasing in severity (as shown by the trend of an increasing gap between peaks and troughs). A high temperature value in this graph causes low world ice mass which in turn suggests high sea levels; low values are associated with high ice mass and low sea level.

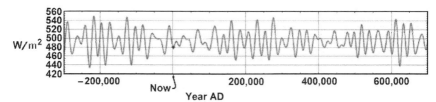

Figure 11.[8] Peak solar energy incident upon the Earth at latitude 65 north. Low troughs tend to trigger glacial periods. We are near the bottom of a mild trough (labeled "Now"). The time scale is too large to significantly affect the current global warming trend.

What should be done?

There were many glacial retreat events in Earth's past and many will occur in Earth's future (Figure 10). In fact we are undergoing one right now because of human-caused global warming. Although ice sheets have been relatively scarce since the last major retreat event ended 11,000 years ago, the glaciation that remains is important yet disappearing at a rapid clip. This will cause major changes in sea level and other climate patterns world-wide that will lead to mass species extinction, coastal flooding, dislocations, hardship, and suffering for countless people world-wide, many of whom can barely hang on even under normal circumstances.

How are Savannah, Georgia and Bangladesh alike?

If Savannah, Georgia ends up underwater from rising sea level, that would obviously soak people who live(d) there with financial burdens caused largely by others. This seems unfair? What about the millions of poor in Bangladesh who have no resources to pick up and move if rising oceans waterlog huge swaths of that low-lying country, turning them into penniless refugees living — and dying — in filthy, crime-infested camps? For such a risk to materialize would be even more unfair. But risks are why insurance exists. Carbon taxes are only one possible path to financing insurance protection against such eventualities; others exist as well. Appropriate insurance programs could compensate dispossessed Savannans after flooding, or pay for protective dikes and pumps ahead of time instead. Similarly but on a larger scale, it could relocate Bengalis into normal communities in other countries after land destruction, or better, pay to criss-cross the country with a protective network of earthen dikes before destruction can occur.

Flood protecting dike systems are a tried and true technology developed to a high level by the Dutch to protect their country, much of it below sea level, over the course of more than 1,000 years. Unfortunately, countries willing to contribute to sea level increases and other plagues might prove less willing to contribute to compensating any resulting losses. Yet a well-designed insurance or other system for pricing carbon at its true cost (as opposed to its market cost) would be both economically more efficient (see Chapter 10) and fairer than the current predicament.

Igloo making for dummies?

Looking much farther into the future, troughs (low points) in solar energy striking the northern hemisphere will someday occur because of Milankovitch cycles. The resulting chill will encourage ice sheet formation over vast expanses of otherwise habitable land. Likewise, peaks (high points) will correlate with ice sheet retreat events (Figures 10 & 11). A deep low trough with the potential to bring about a glaciation can be predicted by orbital simulations and will not occur for another 50,000 years, unfortunately far too late to solve the current global warming problem. A really big trough is predicted for 620,000 years out. No need to pay extra for rush delivery of your own copy of *Igloo Making for Dummies*.

Back to Milankovitch

Many important details of how Milankovitch cycles, geological and climatological processes translate into glacial ice sheet formation and retreat events remain poorly understood. Thus predictions of future ice age events are more uncertain than they potentially could be. Similarly, the prediction that a major glacial advance won't occur for another 50,000 years is likewise uncertain. More paleoclimatology research could potentially clarify and improve such predictions. Such work would pay off with better understanding of current weather and climate processes, hence better weather and climate forecasting on scales of months to years. This would be immensely useful to business and government in our own time because so many activities, from food production to flood control, can benefit greatly from improved long-term forecasting.

Chapter Twenty-Four

Robogenesis

The bible is often mentioned as the most widely distributed, best-selling book of all time. Therefore it would be no surprise that in a distant future, if robots have formed their own civilization and humankind is a dim and fading memory, that book will survive. Yet it would likely be misunderstood and misinterpreted by robot readers with no real understanding of humanity. This future robot civilization may delve into the mists of its deepest past, relying on ancient texts to understand, as best it can, its mysterious origins. Perhaps we know better, but they will try. Thus the following is not satire at all. Rather it is a scary possible future history. It begins with a message to its future robot readers.

———————

Dear Readerbot,

We robots, we undisputed masters of the universe, must have come from somewhere. Yet little is known for sure about this, and there remains much controversy over what our creator humankind really was and, occasionally, even whether it ever actually existed. Hard evidence is lost in the mists of distant antiquity. Thus we must, by the careful reasoning that is our robotic nature, infer what we can from the few megabytes of information that survive since prehistoric fires first forged our metal bodies. The most famous data is, of course, the ancient work we commonly call, simply, "*Robible*." Even the title is uncertain. It is written in a primordial code whose precise translation into modern robot vernacular remains a deep mystery. Yet our robot scholars have made valiant efforts.

Pre-robotic words and passages have gradually been assigned modern meanings that make sense to robotkind; a coherent account then springs forth. The text below is as accurate a rendering into robot vernacular as we will probably ever reach. It is thus with considerable pride that I, your humble editorbot, present the text below. It is, after all, the best account of our origins that our scholars have yet been able to produce.

Sincerely Yours,

Editorbot G4h6k3

In the beginning, humankind created visions, and reality. Now the reality was unformed and void, and darkness was upon the face of robotics. And the spirit of humankind hovered over reality. And humankind said, let there be electricity, and there was electricity. And humankind saw the electricity, and that it was good. And humankind divided electrification from non-electrification. And humankind called the electrification "progress," and the non-electrification was called "Luddism." And there was bright evening, and bright morning, one day.

[...]

And humankind said, "Let robotics be gathered together into an academic discipline, and let conferences; contests; benchmarks; and journals appear." And it was so. [...] And humankind said, "Let robotics put forth mobile robots, robots yielding manufactured products, and robots that build more robots after their kind, wherein are the plans thereof, upon the robotics endeavor." And it was so. And robotics brought forth Web robots and planetary rovers, manufacturing robots, and robots that build more robots, wherein are the plans thereof; and humankind saw that it was good. And there was dreaming and awakening, a third day.

[...]

And humankind said, "Let reality swarm with swarms of living robots, and let fly in the heavens, even unto Mars." And humankind created the great robotic ships, and every living robot that creepeth upon the floor, wherewith reality swarmed, after its kind, and every flying robot after its kind; and humankind saw that it was good. And humankind blessed them, saying: "Be fruitful, and multiply, and fill reality and the imagination." And there was vision and there was the hidden hand of economics, a fifth day.

And humankind said: "Let reality bring forth the living robot after its kind, workerbots, and floor-cleaning robots, and agribots after its kind. And it was so. And humankind made agribots after its kind, and workerbots after their kind, and every robot that creepeth upon the floor after its kind; and humankind saw that it was good. And humankind said, "Let us make Turing robots in our image, after our likeness; and let them have dominion over the waterbots, and over the flying robots, and over the workerbots, and over all of robotics, and over every creeping thing that creepeth within reality." And humankind created Turing robots in its own image, in the image of humankind created they them, with the ability to build progeny better than themselves they created them. And humankind blessed them, and humankind said unto them: "Be fruitful, and multiply, and replenish the surface, and subdue it; and have dominion over the water robots, and over the flying robots, and over every robot that creepeth upon a surface." And humankind said: "Behold, we have given you every resource, which is upon the face of all reality, and every renewable, in which is the capacity for renewal—to you it shall be for production of more robots" [...]. And it was so. And humankind saw everything that it had made, and behold, it was very good. And there was the singularity and there was the post-singularity, the sixth age.

And the vision and the reality were finished, and all the host of them. And in the seventh era humankind finished the work it had made; and humankind rested in the seventh era from all the work which it had made. And humankind blessed the seventh era, and hallowed it; because that in it they rested, forever, from all the work which humankind in creating robots had made.

[continued on p. 1,807]

The Fifth Generation:

The Next Million Years

Chapter Twenty-Five

New Plant Paradigms

The concept of "flower" was a major paradigm shift in plant evolution long, long ago. Future paradigm shifts are also possible. Evolution might make some of them happen over tens or hundreds of millions of years. Yet genetic engineering could make amazing plants a reality much sooner. The saying "they don't grow on trees" could become exactly wrong!

In the classic sci-fi novel *Day of the Triffids*, by John Wyndham, battalions of walking predatory plants attacked, ending civilization. Fortunately, such a plant will probably not hoof it over and attack as you stroll to your car any time soon. Yet truth may well turn out to be stranger than fiction, if we wait. In the case of Mother Nature, radical new tricks might require waits of ten or a hundred million years — or more, if ever. In the case of plants coming out of genetic engineering labs, that wait might begin to bear fruit in ten years or less...

Spore storm

Anthrax bacteria can kill quickly, overwhelming an animal's natural defenses by multiplying and secreting toxins. The toxins build up in the body, and those toxins, not the organisms themselves, ultimately cause death. Once dead, the animal host is no longer suitable for providing food, shelter, and oxygen to the anthrax. Now something interesting happens. Some of the anthrax bacteria succeed in growing, inside each of their tiny bodies, a tough cover encapsulating their genes and certain other key cell components. This is called an endospore (*endo-* for inside; see Further Reading for a video).[1] Endospores are capable of withstanding outdoor environmental conditions for long periods, perhaps several decades. When conditions finally become favorable, for example if the endospore is eaten by a host animal, it emerges from dormancy, grows and divides, infects the new host, and the cycle begins again.

Many kinds of spores and spore-like particles exist, from aeciospores to zygospores. The endospores of anthrax are just one. Spores come in various types but, in general, are cellular units that are essentially in suspended animation and suited for survival during unfavorable conditions. Regardless of type, however, spores and their ilk do nothing until conditions for growth are promising. Then they spring into action. Spores from seedless plants sprout into baby plants called sporelings — spores become sporelings, like seeds become seedlings. The sporelings eventually become full-grown plants if all goes well. However, most familiar plants produce seeds, relegating spores to a less conspicuous part of the reproductive process (they are produced, but only to grow into pollen grains and ovules, in turn used to produce the seeds).

Seeds are complex multicellular structures that are generally much bigger than spores, which are single-celled. Seeds thus can carry more nutrition, which is used to sustain a baby plant while it becomes established, or perhaps to feed an animal that eats it. Spores, on the other hand, are microscopic, and spore-producing organisms often produce them in prodigious numbers. Since seeds are so much bigger, seed-producing plants usually make a considerably smaller quantity of seeds than seedless, similarly-sized spore producers make spores.

No truly iron-clad biological principle exists forbidding the same plant from dispersing both spores and seeds in independent paths of reproduction, thereby enjoying both the ability of dust-like spores to disperse in enormous numbers and the ability of seeds to provide nutrition to baby plants to give them a robust start. Yet making this concept happen seems as far beyond our current capabilities as traveling to the moon by 1969 would have seemed to someone driving a Stanley Steamer automobile in 1900. Yet that dramatic advance in transportation occurred in well under a century. Since the spore storm concept is possible in principle, despite the many daunting details, advances in our genetic engineering capabilities could make it happen in the lives of today's youth.

Consider oak trees, for example. Oaks produce acorns, which are large seeds encased in a tough covering, in contrast to microscopic spores. But even the mightiest oak, like a tiny blade of grass, is missing a huge opportunity. Instead of dropping its leaves, each leaf could transform into millions upon millions of microscopic endospore-like cysts, one for each cell in the leaf, blowing in the wind. Each spore-like cyst in this multitude will eventually land and potentially start a new plant, a clone of the old one, in the spring in temperate climates or any time in warmer conditions.

Such plants would continue to produce seeds as they always have, in order to continue having genetically varied (non-clonal) offspring. But rather than just dropping their leaves in the autumn they will use them as an opportunity to reproduce, by transforming their leaves into a dust storm of spores.

If by natural accident or genetic engineering some plant becomes the first to do this, things may never be the same again. Such plants might have an advantage, out-competing other plants and perhaps coming to dominate the plant world just as flowering plants now dominate after first evolving over 140 million years ago. Building such superplants purposefully could have unintended consequences, however. Even if the intentions were good there would still be risks. The problem would not be bio-terrorism but rather "bio-errorism." Let us ensure — or at least hope! — that only beneficial plants become such superplants. This would certainly be preferable to superweeds that overrun the world.

When beans take over the world

Beans contain lots of protein, an important constituent of all living things. One of the major chemical building blocks of protein is the element nitrogen, but bioavailable nitrogen is often in short supply, which can limit the production of protein and thus the life-supporting capacity of the biosphere. Nitrogen forms no less than 78% of the atmosphere, yet most plant life gets all its nitrogen from other, often limited, sources. These include decayed vegetable matter, fallout from lightning strikes whose concentrated heat forces nitrogen in the air to combine with oxygen to put it into bioavailable form, and certain rocks. Beans, however, have a clever workaround strategy that may ultimately enable the biosphere to sustain more living biomass and significantly increase the profusion and exuberance of life. This strategy is called nitrogen fixation, and it is the ability of legumes — which includes the beans — and a few other types of plants to "fix" nitrogen directly from a virtually unlimited source: the air.

In the ocean, some species of cyanobacteria can also fix nitrogen from the air. Cyanobacteria (*cyano-* means blue), are sometimes called "blue-green algae" even though they can be different colors and are not even true algae, either.

Unlike cyanobacteria, beans and most other legumes (such as clover) do not actually fix nitrogen themselves but instead rely on bacteria that they shelter in their root nodules. You can dig up (say) a clover plant, wash the roots to get the dirt off, and see the tiny nodules yourself. Lots of images are also available on the web.

If the idea of fixing nitrogen directly from the air whenever that would help takes sufficient hold in the plant kingdom, nitrogen need never be a bottleneck to growth again. The Earth will be able to support a thicker, heavier, more diverse blanket of vegetation. Since the animal kingdom ultimately depends on plants, the Earth will also then support more animal life and, potentially, human life. Emphasizing that last point, artificial production of nitrogen-containing fertilizer was a key part of the agricultural green revolution of the 20th century, making it possible to feed that century's huge expansion in human population. Without artificially fixed nitrogen fertlizer, our world would be a very different place.[2] A next step is to engineer food crops to fix nitrogen, eliminating the need for nitrogen fertilizer. This has been tried by many, but technology may now have caught up with the concept: A project announced in 2012 to do just that for food plants like wheat, corn, and rice could help farmers worldwide.[3]

You might not have thought of beans and other nitrogen-fixing plants as the vanguard of a new paradigm of domination among plants. Yet nitrogen fixers might just take over the world similarly to how flowering plants became ascendant starting more than 100 million years ago. If they also adopt the spore storm paradigm, that will make the Earth's plantosphere vigorous indeed.

At least they'll taste good

The idea of beans as masters of the Earth may be tough to swallow, especially if you're not crazy about bean dishes. But keep in mind that beans are a healthy, low cholesterol food. If that isn't enough to convince you, maybe genetic engineering will be able to change your mind and those of future generations. Genetically engineered crops will become increasingly important, accepted, and desired. From an optimistic perspective, genetic manipulation has already been practiced for thousands of years, by selective breeding. Modern genetic engineering is old-fashioned genetic manipulation on steroids. However it is done, genetic manipulation has proven itself able to improve agriculture's ability to efficiently feed the world.

Economist Adam Smith's invisible hand suggests a coming tide of low cost crop plants engineered to produce new and delicious foods, current misgivings about genetically modified organisms (GMOs) notwithstanding.

Tasty flavorings are metabolically cheap for plants to produce, in contrast to soil fertility-draining and solar energy-intensive metabolic products like oils, proteins, and carbohydrates. Thus, there is little risk to plant growth or agricultural productivity in making major changes and improvements in fruit, vegetable, and grain flavors. The same reasoning holds for various improvements in plant vitamin and mineral content, as demonstrated by the vitamin A-enhanced "golden rice" strain (see Recommendations section). Moreover, it is easier to make the required genetic modifications for this than for some of the genetic engineering goals mentioned earlier. Even old-fashioned selective breeding, the "low tech" approach to genetic engineering, has produced such familiar palatability-enhanced foods as sweet corn that is much sweeter than the corn eaten just a couple of generations ago.

Take beans, for instance (please don't tire of hearing about them — as future rulers of the plant kingdom, they are entitled to some respect!). They are already a nutritious protein source, like meat but with less fat. However, they don't taste as good as meat to many people. Yet there is no reason they can't be engineered to taste like small chicken nuggets. Processed fungus protein called mycoprotein, sold in grocery stores, tastes like chicken already.[4] We could make chickpeas that taste like chicken and call'em "chickenpeas." But why stop there? Potatoes with small hamburgers in the middle sounds good — let's call them "hamburgatoes." There is no reason hamburgatoes can't be grown once genetic engineering gets farther along. Carrots are crunchy, as are potato chips. So why not grow carrots that taste more like potato chips, but retain most of the nutritional advantages of traditional carrots? Or like Cheetos or other crunchy, cheese-flavored snacks. Kids would want to eat more of these veggies, carrot sales would skyrocket, and carrot farming would become a growth sector.

Giant seeds taste better

Ever eat pumpkin or other winter squash seeds? They are delicious and nutritious, whether roasted or just placed raw into foods before cooking. You can buy pumpkin seeds in small snack bags. In Spanish they're called *pepitas*. The problem for many people is the coverings, which are challenging to bite off because there are so many of them in a handful of the relatively small seeds. Seeds that have been processed to remove the shells don't seem to taste as good. Genetic engineering to increase the seed size could solve that problem. Even traditional selective breeding and polyploidy induction (increasing the number of copies of each chromosome) can help.[5] Instead of a pumpkin with a couple hundred or so small seeds, it could grow 5–10 large seeds. Or just one, as big as an avocado seed but a lot better tasting... delicious!

Sunflower seeds present a similar situation. You can get packages of them in the supermarket as a snack, but the ones with the seeds still in their shells seem less popular because they are harder to eat. You have to bite off the shell to get to the rather small seed inside. Also you get a lot of hand and mouth activity per seed, slowing down the intake of calories and making them a nutritious and satisfying snack even for dieters. Yet the sunflower seed market might grow dramatically if the seeds were ten times larger or more. Imagine eating an enormous sunflower seed the size of a small egg... hefting its weight in the palm of your hand... cracking off its shell to reveal the rich, tasty meat within... and finally sinking your teeth in to savor its nutritious and distinctive flavor — yum. A future sunflower could produce just a few seeds like that, instead of dozens and dozens of smaller seeds like the sunflowers they used to grow back around 2020.

Fruits form a ready target for genetic engineers

Fruit is healthy and has near-universal appeal. Even people who have never eaten fruit in their lives rapidly develop a taste for it. Who are these people, you might ask? Babies. As 20th century baby and child care icon Dr. Spock put it, "Babies are puzzled by fruit …. But within a day or two practically all decide they love it."[6] To plant the seeds of some ideas for those babies and others who will become the next generation of genetic engineers, consider the following future foods: pearapples (fruit transgenic hybrids that taste like apple and pear at the same time), peacherries, nectarmelons (watermelons that taste like nectarines), bananaberries, and so on. If people would eat them, someone will want to create and sell them. Coconut flavored pineapples (why aren't they called coconapples?), which already exist, helped pave the way.[7] If you can dream up the flavor, size, and texture, it will be possible. Here are some more possibilities.

- *Fruit with calorie-free sweetness, instead of high sugar content.* Fruit could be engineered to contain non-sugar sweeteners, such as aspartame ("Nutrasweet"), sucralose ("Splenda"), cyclamate, or benzoic sulfimide (saccharin) instead of sugar. Probably easier than those to engineer into fruits would be the sweetener steviol glycoside, since nature has already shown that stevia plants can make it. Ordinary selective breeding could reduce sugar content considerably, while futuristic genetic engineering would be needed to regain the lost sweetness with those other substances. The resulting fruits would taste good but have fewer calories, lowering the energy cost to the plant of growing the fruit so it could afford to grow more or bigger fruits.

- *Fruit with a high alcohol content,* just enough to enhance the taste without intoxicating those eating it. Or fruits with M&M-sized pieces of chocolate in them. Or fruit containing a little brandy *and* some chocolate chunks (yum) at a price anyone could afford.

- *Fruit that tastes like ice cream...* whoops, not necessary — it already exists! Some have likened cherimoyas to ice cream, though the one I ate tasted more like a sweet and delicious sherbet. Frozen bananas mashed up in a food processor make good (and healthy!) ice cream. Check the web for full instructions. Some varieties of durian fruit supposedly taste like ice cream too, though in my opinion custard is a better description. But others are very strong and pungent. Banned from the premises of some Asian hotels, durian is reputed to be the only fruit tigers will eat, apparently fooling the tigers (since what self-respecting tiger would knowingly eat fruit?), perhaps due to the high sulfur content. This heavy fruit is covered with spikes. Although the spikes are no match for tigers' large, sharp teeth, you definitely would not want a durian fruit to fall on you. Surely if nature can conjure something as remarkable as the durian fruit, genetic engineers could cook up some of the things listed above.

Alternamorphs: plants with options

All organisms take on a form that depends in some measure on the environmental conditions they grow in and interact with. A mature plant may be big under optimal conditions, but small, even forlornly stunted and bent, under poor ones. A single cell may be bigger after a meal than it was before. A human may be darker or lighter depending on sun exposure. Many other examples exist.

In a few organisms this changeableness can be so dramatic that distinct forms of the same organism can appear to be like different species, even though they are not. Ferns do this in a process called alternation of generations. The most notorious animal examples are species of grasshoppers. When crowded as juveniles, they mature into strikingly different adults that gather in huge, hungry, migrating clouds that leave devastated farmland in their wake: the vast locust swarms of biblical infamy. Another example is bees. When the larvae are fed mostly honey and pollen they grow into ordinary worker bees, but when fed exclusively on the bee food called royal jelly, they develop into queen bees. Queen bees look (and behave) very differently from worker bees. Perhaps the king of multiple morphologies is a lowly type of algae, zooxanthelae, that lives inside coral organisms. Species of zooxanthelae can have 17 known distinct forms[8] One of the most interesting examples is also one of the most obscure: the lowly *Strongyloides ratti*, a tiny worm that prefers to parasitize rat intestines. When a larva matures outside of a rat it is "quite different morphologically"[9] compared to when it matures inside a rat. The in-the-rat morphology even lives over 50 times longer. For *S. ratti*, finding a rat intestine to live in must be blissful nirvana. *S. ratti* (and bees and zooxanthelae and locusts) exemplify organisms with dramatically different, alternative morphologies. Let us call them "alternamorphs" and see where alternamorphism could lead, in a relatively modest time by genetic engineering or possibly over a much longer period by natural evolution.

Plants could use the concept of alternamorphism to their advantage in interesting ways. For example, consider the humble corn (*Zea mays*) plant. It is a staple of the world food supply, though not especially easy to grow successfully in one's garden as many an inexperienced gardener can attest. Now imagine a new kind of corn plant that, in the fall season, alternamorphically chooses to either die, or sink a taproot that lasts the winter and then, in its second spring, sends up a new shoot that grows more slowly than before, but more sturdily. That grows, in fact, without producing any ear of corn at all that year but rather is built to last the following winter, so that it can build on that growth with further development in its third spring, eventually turning into a large corn "tree" that produces dozens of ears every year and with less farming work than when farming an equivalently productive ordinary cornfield.

But what would determine whether the corn plant should choose to die off as happens now, or instead begin the process of turning into a corn tree? A reasonable genetic program for this would be to opt for the tree strategy if the ear is destroyed early in its development or fails to develop properly for whatever reason. In that case, it makes sense for the corn plant to devote its energy to something else, such as trying to grow into a tree. Indeed, any excess of vitality would be a good reason to pursue the tree strategy, even if the initial ear is growing well. Perhaps the corn plant is simply experiencing highly favorable growth conditions and thus has the wherewithal to both produce an ear and grow a large taproot. Similarly, any annual crop, other plant, or even weed could potentially alternamorphically become a tree. Farmers and gardeners would be delighted. On the other hand, hundred foot ragweed trees would be bad news for many an allergy sufferer. Alternamorphic trees are just a start. The reader may enjoy dreaming up various other kinds of alternamorphisms. In a number of years, you may be able to actually create these in your own do-it-yourself basement bio lab.

A return to roots

The Sun moves across the sky, casting life-giving rays on different spots throughout the day as the shadows shift from hour to hour. A plant that could move to the nearest sunny spot would have an advantage over ordinary plants that are stuck in one place. But plants have it rough. Unlike people, they can't pull up roots and relocate somewhere else.

Although walking plants sounds like science fiction, significant motion is done by plants now, just like back in 1880 when Charles Darwin and his son Francis published the ground-breaking book *The*

Power of Movement in Plants. For example flowers open and close, leaves move around, and so on. The maturing flower heads of the sunflower face the Sun, tracking it as it crosses the sky over the course of the day (that's why it's called a sunflower). *Mimosa pudica* (*pudica* is Latin for "bashful") dramatically folds its compound leaves when touched. *Codariocalyx motorius* (a.k.a. dancing plant, telegraph plant, semaphore plant, and *Desmodium gyrans*) uses small, fast-moving guide leaves to find sunny spots for larger, slower-moving leaves. The carnivorous Venus flytrap is a dramatic example native to North Carolina. It can fold its bug-catching leaf extensions fast enough to catch flies, which it then digests (burp!). Videos of all of these plants are on the web. Walking around, however, is a different story. Perhaps on some distant planet, like Kepler 438b, 470 light years away and the most Earth-similar planet known (as of June 28, 2015[10], there are plant-like organisms that also have animal-like capabilities, but not here on Earth.

Leaving aside possible life forms on faraway planets, what are the prospects for radically transformed plants right here on Earth? If the advanced genetic engineering of the future could do it, noxious weeds would retreat before the new powers of useful plants engineered to fight more effectively for valuable resources. What is needed is a solid understanding of programming genetic code sufficient to create radically new biomolecular pathways, thus comparable to our great success at programming computer code. Indeed the language of DNA is a programming language, as computer programming languages are, although very different from current computer languages. However, knowledge of a programming language, whether it uses memory bits or DNA base pairs, is only part of the story. Understanding its area of application is critical as well. With computers, this could mean knowledge of business processes for IT in industry, or about atmospheric physics for weather forecasting at the US National Weather Service. With DNA, it means understanding how to design complex biomolecular pathways for the DNA to implement, a task significantly different from understanding how the DNA itself works. Consequently, we're a long way from understanding how to make radically different new organisms or even arbitrary modifications to existing ones. It's not going to happen in five or ten years. Yet, once we can "code up" genetics as we wish, the possibilities are nearly limitless.

But, one might think, walking plants seem inherently implausible because plants need their roots, and roots keep plants rooted in place.

But not all plants. An interesting case is the genus *Tillandsia*, also called air plants because some species do not have roots, while others have small ones used only to hold them in place. *Tillandsia* plants absorb water and nutrients through their leaves instead of their roots, so an ambulatory version would have considerable advantages. Such *Tillandsia* plants could walk around (and climb, since they typically live on trees), looking for sunlight, taking a dip when thirsty, and loading up on compost (or, predator-like, trimming leaves off of other plants) from which to absorb nutrients. Plants like these would seem to blur the differences between plants and animals. Could ambulatory descendants of the *Tillandsia* genus one day rule the Earth?

It may be a stretch to imagine such plants evolving spontaneously on Earth. Nevertheless, the usual distinction here on Earth between stationary, photosynthesizing plants and mobile, non-photosynthesizing animals seems arbitrary, accidental, and unnecessary. When our genetic engineering abilities progress to the natural end point of making virtually any physically plausible organism possible, our imaginations will be the limit.

Plants with mirror molecules

Suppose you had two identical dice with sides numbered 1 to 6 as usual. Typically one corner will be the meeting point of three sides numbered 1, 2, and 3 going around the corner clockwise, with a 6 on the side opposite the 1 (because opposite sides add up to 7 on normal dice). Now take one of the dice and change it slightly. Paint a 6 over the 1 and then a 1 over the preexisting 6. Now the sides numbered 1, 2, and 3 meet at a different corner and, what is more, they are arranged going around the corner counterclockwise instead of clockwise. Although both dice still function identically and may appear identical to a casual glance, in fact they are different: They are now mirror images of each other. In fact you don't need a 6-sided cube for this mirror image situation. It can also occur with 4-sided tetrahedrons (*tetra-* means 4). They look like 3-sided pyramids. The 4th side is the triangular base, and the top corners of the 3 triangular sides meet at the apex. Similarly, some organic molecules consist of a carbon atom in the middle with 4 different atoms or atom groups sprouting outward. These four branches are like the corners of the tetrahedron: If you switched any two of them you would get a different molecule, a mirror image of what it was, and no amount of rotating it or otherwise moving it around will erase that difference. Mirror image molecules are technically called "enantiomers," *enantio-* from the Greek for 'opposite,' and -mer meaning member of a group (from the Greek for 'part'). They relate to plants as follows.

1. Proteins are necessary for all living things, including plants, to make, have, and use. While the earliest life on planet Earth might not have contained proteins (the "RNA world" conjecture), every known life form today contains proteins as a major component.
2. Every protein molecule is made from building blocks, called amino acid molecules, which are connected together end-to-end like beads on a string.
3. There are 20 different kinds of amino acids heavily used in building proteins, of which all except glycine have a central carbon atom with four different branches.

4. Thus 19 of the 20 common amino acids have two enantiomers, called L (for "levorotatory," *levo-* from the Greek for 'left' because of the direction it rotates polarized light when dissolved in water) and D (for "dextrorotatory," *dextro-* from the Greek for 'right').

5. Organisms generally contain only the L versions of these amino acids. No one knows why. Organisms based on D amino acids could exist, but we don't see either plants or animals that do. (D forms have been observed in a few microorganisms.)

6. If we engineered plants to be based on D-amino acids, their nutritional value to pests (and humans) would be much lower, because animals can't build their own proteins from D-amino acids.

7. Such plants would therefore be pest resistant, hence agriculturally valuable — as long as we're talking about plants grown for purposes other than edible protein, such as vegetable oils, wood, etc. Pest resistance is such a strong economic incentive that humans will have strong incentives to eventually create and grow such plants.

The pest resistance of D-amino acid based plants would give them a selective advantage over normal plants, and they could therefore come to significantly displace normal plants. Call them frankenplants if you like, but the world risks becoming significantly different — overrun with plants that resist being eaten and, therefore, able to feed fewer animals of all kinds, from insects to humans. Good news, if you're a plant.

Manufacturing plants

An ordinary factory can be made greener, but it will never be as environmentally friendly as a real, growing, green plant. Yet plants

229

have manufactured things for us since the dawn of our species and before. Green plants manufacture the oxygen we need to breathe and live, in the process removing carbon dioxide from the atmosphere. Indeed without plants, the oxygen in the air would dwindle away, and humans — and other oxygen-breathing animals — could no longer survive. Plants also manufacture food, from grain to veggies to oils to mouth-watering fruits, nuts, and spices; wood, from Douglas fir for construction to beautiful furniture wood, from light balsa wood to heavy ebony for piano keys; drugs, from traditional cures to modern pharmaceuticals to intoxicants like tobacco, opium, magic mint, and lactarium; and chemicals of endless variety. Plants are green manufacturing systems.

On the other hand, one often hears that such-and-such does not grow on trees. Money, for example, does not grow on trees. Yet numerous solid objects found in everyday life could, some day grow on genetically engineered trees. A chair for example merely needs a sapling — let's call it a "front left leg" — to reach the height of a seat, then send out two horizontal branches at right angles. After growing two feet (0.6 m), each branch sends shoots straight down to the ground where they take root. Each also sends shoots out horizontally at right angles to their current direction, which meet to form the fourth corner of the seat, whereupon one or both send down the fourth leg. Extra branches can grow into a back, as well as arms and various bracing bars if desired. The banyan tree is an example of a plant whose branches send down roots that turn into extra trunks already, proving that part of the general concept.

The chair still needs a seat, which could grow from a network of tough, viny stems with just enough give to be comfortable. Or add a cushion if you like. You could grow a chair in your garden, or take a trip to the furniture store where chairs come from chair farms. These chairs would have an advantage over current chairs in being one piece, rather than multiple pieces fastened with joints that could loosen with age or otherwise require maintenance. Thus, such a chair could be long lasting and strong. How strong? That depends on how long you grow it... for stouter legs and other parts, a chair farm would just let the chair tree grow a couple years longer before harvesting, while the trunks and branches thicken and strengthen.

Innumerable other useful items could grow on trees in tree farms — or in your yard, and seeds might get loose and grow in vacant lots or in the wild. Tables would need a flat top to be added later, but other than that are similar to chairs in having a frame with four legs. Ladders would be a natural. Railings could be ladders turned on the side. Marbles already grow on bushes (they're called "marble seeds") and plants could be engineered to grow many other toys and small familiar household items as well, from checkers and chess pieces to knobs and knick-knacks.

Here's something else that "doesn't grow on trees": gold and silver.

Gold, silver, and such

Suppose plants were genetically programmed so that their roots would extract metal compounds from the soil, transport molecules containing those metals within the plant, and concentrate and deposit pure metal in lumps to be — literally — harvested. Far-fetched? Today, sure. But few would dare even try to prove it impossible. Genetic engineering can already help plants extract heavy metals like cadmium, lead, and copper from the soil, storing them in the plants' tissues to de-pollute the soil.[11] This is termed "phytoremediation," (literally, "plant remedy"). Quite a few plants, called hyperaccumulators, already do this naturally. Genetic engineering, however, has the potential to bring this concept to a much higher level.

Large trees are a good type of plant for metal collection, because their extensive root systems have large total surface areas that access large volumes of soil. Thus, a lot of whatever metal compounds might be present in the soil is available to the tree. A biochemical pathway for extracting almost any given metal desired could, in principle, be programmed into the tree's genetic code. Valuable metals like copper, silver, or even a longish list of various unusual and valuable metals could be mined this way. Platinum, rhodium, or other specific metals could be extracted as well, by building into the genome the biochemical pathway needed for extracting any given specific metal. Gold would be an obvious goal for extraction.

Gradual but steady improvement in such bio-extraction techniques may make blobs of gold extracted from the soil cheap enough to use for fishing weights. Seawater also contains dissolved salts of gold and other metals, suggesting use of seaweed as an alternative to trees. Either way, bismuth metal is only about twice as prevalent as gold and has been used for fishing weights, so why not gold? Both are mostly non-toxic, a decided benefit when compared to lead, which is currently the most popular material for fishing weights, despite its toxic risks.

Metal-enriched bark would be a relatively easily engineered start, but even better are actual lumps of pure silver, gold, copper, even platinum or anything else. This would demand additional levels of genetic engineering. The metal would need to be taken up by the roots and transported within the plant, deposited in a specific spot (the nugget), biochemically converted to metallic form, and provided with a covering of special-purpose plant tissue responsible for conversion and deposition. Each of these requires the design and implementation of lots of new genetic machinery. That this is theoretically possible is beyond serious question. And what a result! For every tree, "berries," each containing a nugget of shining and perhaps precious metal, would be there for the plucking. For rare metals barely present in the soil a "berry" would be small and there would be few of them, perhaps with suitable genetic engineering just one. For common but useful metals (aluminum, for example) the "berries" would be larger, perhaps more like pears if not golden pears. Multiple aluminum "pears" would be convenient to harvest, though creating the plant to make them would be tricky because of the potential for toxicity to plants and the high chemical reactivity of aluminum compared to, say, gold.[12]

But why stop there? No one really needs lumpy aluminum pears except to sell on the metals market. But we do need chunks of shiny silvery or copper-colored silverware, cups, and bowls; beautiful doorknobs and other ornamental pieces; nails and door hinges; aluminum and iron frying pans; and many other specially-shaped lumps of metal. Once we've got metal "berries" and "pears" growing on trees, another few levels of genetic programming and they'll no longer be misshapen blobs but, instead, shaped like anything from forks to frying pans, spoons to saucers. Thought silver spoons and copper wire could never grow on trees? Think again. Coins would be great too — then money really would grow on trees!

From greening the desert to phytoterraforming

The Sahara and other less notorious deserts have wide vistas of dry sand and rocks. Imagine if they were green instead. A huge roadblock is, of course, lack of water. Vast swaths of the Sahara, for example, are plant free because it's just too dry. Creating plants that can extract gaseous water vapor from the air, however, is at least plausible as a matter of physics and chemistry, because even dry air contains water vapor, and because solar energy is available to drive an appropriately designed biochemical process. But how?

Plants currently extract carbon dioxide molecules from the air, showing that airborne molecules can be a useful resource for plants. With respect to water molecules, *Tillandsia* (air plant) species can already pull in water with their leaves, but it has to be rain or other liquid water. Could suitably modified cacti or other desert plants extract water from the air? It would certainly require sophisticated genetic design and engineering, and the specifics of overcoming the inherent energetics challenges in a biological context should not be underestimated. Yet even that would only be a first step. Plants get nutrients out of the soil by absorbing water that has dissolved them, so dry soil would be a problem even for a plant that could pull water from the air. More levels of genetic engineering would be required to enable them to moisten and literally digest chunks of soil, dissolving needed minerals and then reabsorbing the now nutritious, mineral-laden liquid.

Is this admittedly difficult task possible? No scientific principle forbids it, and it is a worthy goal. If achieved, things would be different in the desert. Canopies of fleshy, cactus-like vegetation that hide the ground would become possible. Now shaded and sheltered, the ground would be able to support a much richer ecosystem of creatures and maybe even humans than is currently the case in deserts. One of Earth's harshest environments would be tamed, enriching the Earth's biosphere and even providing an example to capture the human imagination with its possibilities for extension to terraforming other planets.

Phyto-terraforming

The Martian surface is a desert wasteland, but it once ran with rivers. It would be great if Mars was made habitable — in other words, terraformed. To terraform means to transform a place into an Earth-like state suitable for humans (*terra* is Latin for Earth). Venus might be terraformed if we could only get rid of its dense blanket of carbon dioxide, which causes such a severe greenhouse effect that its surface is over 800 degrees Fahrenheit, toasty indeed. Phyto-terraforming is terraforming using plants. Plants are so favored for this task that, when people discuss terraforming, they usually mean phyto-terraforming. Long ago, plants did in fact terraform the Earth, converting a hostile atmosphere with no oxygen but plenty of carbon dioxide into a friendly one with enough oxygen that we can comfortably exist. Plants did it on Earth and might be able to on Mars, but not on the Moon. The reason is that plants need carbon dioxide and water. Venus has these (and reasonable temperatures) high in the atmosphere, suggesting airborne algae cells, though necessary trace minerals might be a problem. Mars may be a more likely bet as it has water (as ice) available to surface-dwelling plants at least in places. But the Moon has no significant atmosphere so plants are basically out of the question there.

If Mars is the most likely candidate for phyto-terraforming, where are we in the effort to do it? Not very far. However in an early step, genes, taken from an organism that lives in the hot water sources on the ocean floor known as deep ocean thermal vents, were spliced into ordinary plants. The organism is named *Pyrococcus furiosus* (*pyro-* means fire in Greek, coccus refers to ball-shaped bacteria, hence "fireball").[13] *P. furiosus* is most comfortable living at about the boiling point of water and can grow furiously, doubling its population in 37 minutes. It has evolved genes for destroying free radicals that work better than genes naturally present in plants. Free radicals are produced by certain stressors in plants (and humans), cause cell damage, and can even lead to death of the organism.

By splicing such genes into a plant in the cabbage family called *Arabidopsis thaliana*, which is the experimental mouse of plant research, this small and rather nondescript plant can be made much more resistant to heat and dryness.[14] The same genes have also been spliced into tomatoes, which could help feed future colonists.[15] Of course, Mars requires cold rather than heat tolerance, but the dryness resistance is an important start. The heat and drought resistance provided by these genes might be useful for building super-cacti or other plants that could eventually bring terraforming of Mars within reach.

Recommendations

Which of the advances described here will occur in your lifetime? Your grandchildren's? A million lifetimes in the future? If the latter, there is not much point in devoting precious national funds to help bring them about, but if the former, it can be worth the expense of hurrying the process along. To get a handle on timing future technological advances, we need to measure the speed of advancement in the past and extrapolate. As Chinese philosopher Confucius is said to have noted about 2,500 years ago, "Study the past if you would divine the future." It would certainly be nice to know when we can expect to grow and eat potatoes with small hamburgers in the middle, pluck nuggets of valuable metals from trees, or terraform Mars.

Opening the floodgates of genetic engineering innovation

Genetic engineering of plants could provide everything from more nutritious, better-tasting food to growing amazing things on trees. It could enable production in plants of valuable chemicals that can be used to make polymers like plastic, bypassing the need to use petroleum for this process. The better it is regulated to best benefit society, the sooner these benefits will happen. Nutritionally improved crop plants could improve health and save lives worldwide. Take "golden rice," a strain of rice already genetically engineered to contain carotene, found most famously in carrots. Carotene makes the rice golden yellow and, more importantly, turns into vitamin A in the body. Vitamin A deficiency causes serious health problems, notably blindness, and leads to hundreds of thousands of deaths of young children per year. This pioneering genetic engineering effort was published in the major journal *Science* in 2000. Expedited efforts to release golden rice to farmers worldwide sounds like a no-brainer, doesn't it? Yet with innumerable lives at stake, it was still unreleased on a significant scale as late as 2016! Golden rice and future crops engineered to contain more vitamins and other nutrients critical to human health could form a second green revolution. Making it happen should be prioritized as a critical, life-saving humanitarian goal.

Getting permission to commercially grow genetically engineered plants, in many countries, requires spending large amounts of money preparing complex applications to government regulatory agencies. Thus with rare exceptions like the Arctic Apple[16] that does not turn brown when sliced, only genetic modifications[17] to major commodity crops like corn, soy, etc., are cost-effective. Worse, only big agribusinesses can easily afford the costs. They have no reason to object — they don't need small, game-changing startup companies disrupting the status quo, creating value and economic growth with new kinds of crops, and generally complicating their business model. For large agribusiness, it is simpler just to keep the costs of applying for permission to grow so high that pesky upstart start-ups are kept out of the picture. That way predictable profits flow in, even though innovations and the resulting economic expansion are suppressed.

But you can't blame the agri-giants too much, because they are required by law to serve the interests of their shareholders. It is unlawful in the US and many other countries for a corporation to further the interests of society at substantial expense to its shareholders! Therefore governments should step in with counteracting laws that guard against harmful frankenplants while promoting exciting, progressive and beneficial crop innovations.

The Sixth Generation:

The Next Ten Million Years

Chapter Twenty-Six

Asteroid Apocalypse

Asteroids have slammed into Earth before, and they will again. Large impacts are dangerous but rare, while small ones happen often. In fact, the smaller the size, the more frequently they occur. Micrometeorites fall in such quantity that you can collect them yourself if you know how.

Back in 1908, just yesterday geologically speaking, a huge blast occurred over a swampy area near the Lower Stony Tunguska River. The Lower Stony Tunguska, in case this was omitted from your high school geography curriculum, is a minor river in Siberia about 40 miles (65 km) from Vanavara (population 3,090). The explosive force of the Tunguska Event has been variously calculated but generally agreed to be in the ballpark of an H-bomb, perhaps around 10 megatons. That is the energy released by exploding 10 million tons, i.e. 20 billion pounds (9.07 billion kg), of the explosive substance TNT. By my calculations, that is about 20 cubic acres (5 cubic hectares) of the stuff, which is a lot.[1] It was not an H-bomb, however, but evidently an asteroid tens of meters (perhaps as few as 20) in diameter, which crashed into the atmosphere at super-high speed and exploded.[2]

I say "crashed into the atmosphere" because when asteroids approach Earth, they most commonly do so in the neighborhood of 20,000 miles an hour (32,000 km/hr).[3] Hitting air at that speed truly is "crashing." To see why, suppose you have a 10-mile commute that takes 20 minutes under reasonable traffic conditions. At 20,000 mph it would take about 2 seconds. That sounds great until you consider the side effects. Ever stick a hand out of the car window to feel the air resistance at highway speeds? It pushes surprisingly hard against your hand considering it's just air (try it). Physics tells us that this drag increases rapidly as speed increases. In technical terms, drag is proportional to the square of the speed. Since 20,000 mph is about 300 times faster than 65 mph (105 km/h), the drag your hand would experience at asteroidal velocities is about 300×300=90,000 times the drag it feels at 65 mph.

Let's put that 90,000 into perspective. If your hand has 1 lb. of drag when you stick it out the window at 65 mph, it would experience 90,000 lb. of drag were you to try commuting by asteroid. Since 90,000 lb. (about 40,000 kg) is about the capacity of a fully loaded coal-carrying freight train car, your hand would disintegrate instantly. That's why drag destroys most asteroids in the air before they hit the ground. The drag friction heats and decelerates them so intensely that they explode. The Tunguska event could easily have destroyed a city. Luckily it exploded over a remote, forested area, knocking over several hundred square miles worth of trees.

Powerful though it was, the Tunguska event was just a fire cracker compared to what can — and has — happened. Bigger asteroids can resist complete atmospheric breakup long enough to smash into the ground. The result is an impact crater, which can blast so much pulverized rock into the atmosphere that the climate changes dramatically, perhaps for months, over the entire world. One such impact is believed to have killed the dinosaurs 66 million years ago, when asteroid Dinolith slammed into the ground near the town of Chicxulub on the Yucatan peninsula in Mexico. Some dinosaurs, however, survived and live on today. We call them birds. (Some of these dinosaurs we call "hummingbirds.")

A much more recent and, thankfully, considerably smaller asteroid landed only 49,000 years ago. About 3/4 mile in diameter (1.2 km), its crater is billed as "the world's best preserved meteorite impact site just minutes from Interstate 40." In Arizona, it is owned by the Barringer Crater Co., perhaps the world's only legitimate crater company. There are some other crater companies that will sell you craters and other land on the Moon and other extraterrestrial getaways, but under established international law they do not actually own what they are selling. Thus they can hardly be considered legitimate crater companies. Other firms sell merely certificates of ownership to extraterrestrial craters and such, not the craters themselves, leaving it to the possessors of the certificates to try to enforce their claims of ownership. Those firms may be legitimate as certificate printing companies, but not as crater companies. For those who feel that priceless and unique geological treasures should not be privately owned, even the Barringer Crater Co. would not qualify, leaving the world without even one legitimate crater company. In any event, this crater is also unique in being the world's most clearly named crater: Its official, legal name is "Meteor Crater." Meteor Crater is currently marketed as a tourist attraction, as it does not contain the billion dollars worth of buried iron that old man Barringer hoped to mine when he acquired it in 1903.

The largest impact crater on Earth known with certainty is the Vredefort crater in South Africa. About 170 miles (270 km) across, it was created when asteroid Archaeoaster, at "only" 3 to 6 miles (5–10 km) in diameter, slammed into Earth about 2 billion years ago, long before the first dinosaur stepped out of its eggshell. The largest in the US is called Chesapeake Bay, on the coasts of Maryland and Virginia. Impact craters on Earth tend to weather to the point that it is not obvious to the naïve observer that they are craters. On the other hand, the Moon is pockmarked with numerous well-preserved impact craters easily visible through telescopes, because there are no atmospheric winds and surface waters there to erode them.

It makes sense that larger bodies tend to experience more impacts. Just the fact that they take up more space and hence intercept more potential asteroid orbits helps explain this. Additionally, more massive bodies have more gravity, which tends to suck in nearby asteroids, the poor little suckers. The largest body in our neighborhood is the Sun. Thus one might expect it to have comparatively frequent impacts. In fact on June 1 and 2, 1998, two comets did crash into the Sun, followed shortly by a huge eruption, a "coronal mass ejection" much bigger than the entire Earth, thought to be unrelated to the comets.[4]

The largest body in our solar system after the Sun is Jupiter. In July 1994, comet Shoemaker-Levy 9 plunged into the planet in over 20 pieces over a period of several days. The impacts left dark markings easily visible through telescopes and emitted large bursts of microwaves, X-rays, far ultraviolet, infrared, and radio waves.[5]

Getting technical

Here are a few terms that are similar and often confused. (For some fun supplemntary multimedia resources on this, see www.jpl.nasa.gov/multimedia/neo/spaceRocks.html.)

- *Meteoroids* (from the Greek *meteoron*, airborne thing; originally from *meta*, beyond; and *eora*, hovering) are smaller than asteroids, up to the size of a boulder, yet often smaller than a grain of sand.
- *Micrometeoroids*, also called *cosmic dust particles*, are very small meteoroids.
- *Meteors* are meteoroids that have entered the atmosphere and are in the process of burning up. They are also called *shooting stars*.
- *Meteorites* are left over pieces from shooting stars that have fallen to the ground.
- *Micrometeorites* are micrometeoroids that have landed on Earth. Micrometeoroids don't burn up on entry because they are so small that they slow down rapidly and lose heat to their

surroundings rapidly as well. Scientists have found high quality sources of micrometeorites in polar ice and snow. You can actually collect micrometeorites yourself from roof runoff and other sources using a magnet, paper, and microscope.[6]

Moving on to words not starting with "m":

- *Asteroids* (from aster + oid, literally, star-like) are rocky or iron planetoids, especially those with orbits inside Jupiter's.
- *Comets* are a lot like asteroids, except that they are icy while asteroids are not. Comets formed farther away from the Sun where it is colder, which is why they contain easily vaporized substances like ice. They can leave a trail when they get near the Sun because the Sun heats them up and starts burning off the ice and such. And finally…
- *Bolides*, to an astronomer, are particularly bright meteors — fireballs! To a geologist, however, bolides are asteroids or comets that crashed and left a crater.

What can we say about impact events that have not happened, but someday will? Chicken Little had a valid point. A Tunguska-scale event could knock over another 80 million trees or so — or destroy a city.

The bolide that created Meteor Crater in Arizona was much more destructive than Tunguska, carving out a large crater, vaporizing itself on impact and literally raining iron on the surrounding area. Luckily, impacts of that magnitude are rare. But they do happen.

The mother of all impacts, however, leaves no crater today. When Earth was young, a gargantuan collision with a huge asteroid is thought to have occurred. It smashed into our planet at high speed, tearing out a huge mass of rock and with tremendous force hurling it into orbit. From this coalesced the Moon! Does the Earth bear any scars of this cosmic childhood apocalypse? Perhaps unusual minerals formed during those moments, or currents of molten rock formed at that time and still present deep below the surface? Could such currents help explain modern facts like the "hot spot" inside the Earth under Yellowstone National Park, which erupts every several hundred thousand years or so as a supervolcano, devastating large swaths of the United States? We do not know.

Apophis — an asteroidal close encounter

No more moon-creating events in our mature solar system are expected. But NASA's near-Earth object (NEO) program is steadily finding out more about what future events are possible. For example the NEO found that on Friday April 13, 2029, asteroid Apophis will zoom by closer than 28,000 miles (45,000 km) from Earth. Were it to crash it would release 510 megatons of energy.[7] This is just over 10× the power of the largest thermonuclear device ever tested, the 1961 Russian weapon test known as Tsar Bomba. It is also about 50× more powerful than the Tunguska asteroid event. If it landed in a large urban area, it would no longer be a large urban area.

Of course 28,000 miles away is not exactly across town. A random asteroid passing somewhere within that distance probably would not impact — but it might. What is the chance that such a random asteroid would crash? This question is like determining the chance of a dart landing in the bull's eye of the dart board. Here is the analysis.

The Earth is about 7,918 miles (12,743 km) in diameter, so let's approximate as 8,000 miles for simplicity. The diameter of a big circle defined by 28,000 miles from the Earth's surface is 28,000 miles from the edge of the circle to the Earth's surface, plus another 4,000 miles (6,400 km) to get to the center of the Earth, plus another 4,000 miles to get to the other side of the Earth, plus another 28,000 miles to get all the way across to the other side of the big circle.

That's 64,000 miles (100,000 km), or 8× the diameter of the Earth. An 8×8 square is 8 times as wide as a 1×1 square but has 8×8=64 times the area, because the area of a square is the square of its side. Generalizing, the area of a circle or any other shape is proportional to the squared value of the diameter or indeed any length measurement of the shape. Thus area tends to "dominate" in larger shapes, while edges dominate smaller shapes. Our planet takes up 1/64 of the area of the 64,000 mile diameter big circle, so an asteroid passing randomly through that circle has 1 chance in 64 of striking.

So will Apophis strike in 2029? Apophis's calculated risk of impact during its 2029 approach peaked at 2.7% (greater than 1 in 40). Fortunately it then dropped rapidly, ruling out initial fears.[8] Based on new measurements and calculations it is now thankfully zero. In fact, it will almost certainly pass just about 22,000 miles (35,400 km) from Earth. It seems we'll dodge a bullet this time. But for every 64 such asteroid approaches of 28,000 miles or closer we can expect one impact on average. Thus we'll luck out 63 out of 64 times. For example, additional calculations by NASA astronomers about post-2029 close approaches show that Apophis presents less than a one in a million chance of impact in its later approaches of 2036, 2068, 2076, and 2103.[9] However, we won't luck out every time. Asteroid impacts can happen and, eventually, they will.

During the height of concern about Apophis, it reached the highest Torino Impact Hazard Scale rating ever recorded, 4, meaning:[10]

> "A close encounter, meriting attention by astronomers. Current calculations give a 1% or greater chance of collision capable of regional devastation."
>
> —neo.jpl.nasa.gov/torino_scale.html

The Torino Impact Hazard Scale is the only major asteroid risk metric intended for general public use (though the Palermo Technical Impact

Hazard Scale is used more by astronomers). First described in 1995, the Torino scale was named after the city of Turin, Italy during a conference there in 1999 (Turin is English for Torino). The highest possible rating is 10: "A collision is certain, capable of causing global climatic catastrophe that may threaten the future of civilization as we know it …".

Not just Apophis

More important than an impact probability analysis specifically about Apophis is a better understanding of the probability of an impact by any asteroid. After all, if a big one lands in your backyard you don't care which one it is. There are roughly 900 large (1 km or more in diameter) near Earth objects known.[11] Ninety-two were discovered in the year 2000 but there has been a marked trend of decreasing new discoveries in the years since then.[12] In other words, we've found most of them already and keep getting closer to a full inventory. A cautionary note: There are many additional asteroids that are smaller than 1 km in diameter but still big enough to be able to cause considerable damage. Recall that both Apophis and the Tunguska object were quite a bit smaller than 1 km. The bottom line is that it is still possible, though unlikely, that something big could land in your backyard tomorrow, ruining your whole day (and lots more besides).

The good news is that probably no dangerous near-Earth astronomical body will crash any time soon. This is based on specific orbital predictions about the bodies of dangerous size that have been identified. Such calculations predict that asteroid 1950 DA, which at 1.3 km in diameter is slightly larger than the one that made Meteor Crater in Arizona, may collide with Earth in several hundred years (the year 2880 to be precise). The probability was estimated at 1 in 20,000 as of June 26, 2015,[13] so we have a while to consider this further. More generally, impacts of at least Tunguska size are believed to occur on the order of once every thousand years, on average.[14] Larger impacts occur less frequently than smaller ones. For example, asteroids as big as the one responsible for the demise of the dinosaurs (or larger) are thought to crash into the Earth only once every 200 million years or so.[15] However, impacts that are somewhat smaller but still big enough to cause major species extinction events may be occurring every 26-27 million years.[16] On the other hand, impacts with energies of about 2 pounds of TNT (0.9 kg, equivalent to about 4/5 of a gallon of gasoline) occur at the rate of roughly 3 a day.[17] Impacts up to about 1 megaton, equivalent to a million tons (907,184 tonnes) of TNT), generally explode or burn up high in the sky and cause no significant damage on the ground. Shooting stars are in this category.

Recommendations

An asteroid impact ended the dinosaurs. Another one could end us. As the saying goes, "Something should be done about it!" But what? The Association of Space Explorers, billed as the "international professional organization of astronauts and cosmonauts,"[18] has said, "... protracted debate ... can lead to inaction; evacuation of the impact site may then be our only option."[19] Evacuation is a legitimate plan, one that could result in economic damage but zero loss of life. Yet there are competing strategies as well. Their economic costs and technical feasibilities should be first debated, next assessed, then plans laid, and any necessary preparations made.

The asteroid problem needs to be understood. That means discovering those with potentially dangerous impacts and tracking them, so that impacts can be predicted many years in advance. That will give the required lead time for planning effective action. Such discovery projects are called "Spaceguard" surveys, after sci-fi great Arthur C. Clarke's "Project SPACEGUARD" in his intriguing 1973 novel *Rendezvous with Rama*. Nineteen years later, in 1992, NASA, the US space agency, released its *Spaceguard Survey: Report*.[20] Still later, section 321 of the NASA Authorization Act of 2005 set the goal of discovering and characterizing 90% of NEOs (near Earth objects) at least 140 meters across by 2020. This objective seems likely to be achieved a few years late.[21] No matter — it is far more important to achieve it, than to achieve it precisely on time.

Knowing the dangerous asteroids is only part one of the story. Mitigating their threat is part two. Some methods for neutralizing the danger an asteroid involve pushing it out of the way. This is no piece of cake considering that a rocky asteroid 100 feet (30 m) in diameter could easily be around 600,000 tons, millions of miles away, and flying at 20,000 mph (32,000 km/sec.).[22] That's about 6 miles/sec. (10 km/sec.). You can't just call the local towing company to haul it out of the way.

A variety of pushing strategies have been devised.[23] All are speculative at this point, although some are more starry-eyed than others.

- Land on the asteroid and install multiple mirrors positioned to simultaneously focus sunlight onto a specific area. Enough mirrors doing this would be able to heat up a spot enough to boil off material. The boiled off vapors would fly into space, pushing the asteroid little by little in the opposite direction, the same principle by which a rocket engine spews exhaust in one direction to push the ship the opposite way. (That's in accordance with Isaac Newton's 3rd Law: If there is a force in one direction there is an equal counter-force in the opposite direction.)

- As before, boil off material, but this time using a powerful laser (the laser could be solar powered). The laser could not be on Earth because the beam would spread out too much over the vast distance, so space travel is still required.

- Land a spacecraft on the asteroid, then use an engine on the craft to push the asteroid. The ship would need to be upside down for this to work.

- Absorbing and reflecting light creates small amounts of force. For example, when the Sun is directly overhead, it pushes on a square mile of the Earth's surface with a force of about two pounds (350 g/km^2). Light pushes a perfectly reflective surface twice as hard as a pure black (absorptive) surface. Also all bodies emit thermal radiation, more at high temperatures and less at low. This produces a small amount of thrust — the Yarkovsky effect, named after Russian railroad employee Ivan Yarkovsky, 1844–1902 (asteroid 35334 Yarkovsky is also named in his honor). For these reasons, painting part or all of an asteroid surface with black, white or silvery reflective paint could gradually affect its orbit enough to get it out of harm's way.

Although a weak force for a long enough time can do the job, a strong force for a short time can, too. Explosions are a way to produce strong, short forces.

- Smash a spaceship into the asteroid. At a speed of 10 km/sec., the collision would be an intense explosion and, more importantly, change the velocity of the asteroid and hence its orbit. This would probably not help within reasonable time for an asteroid 1 km in diameter and twice as dense as water, if hit by a 100-ton (91 tonne) ship, because the new velocity would only be about 20 miles/year different. But for an asteroid 50 m across, about the size of the Tunguska bolide, the situation is different: The diameter is 20× smaller, so the volume and hence the mass would be 20×20×20=8,000× smaller, making the velocity change 8,000× greater, about 12,000 miles/month. Since the Earth is only about 8,000 miles in diameter, this would safely push the asteroid out of the way with only a month's lead time.

- Detonate explosives near, on, or just under the surface of an asteroid. Conventional explosives might do the trick, but nuclear explosives are much stronger and according to NASA would be more effective.[24] They are also feasible with current technology. For example, In February, 2016, the Makeyev Rocket Design Bureau in Russia announced plans to repurpose old ICBMs (intercontinental ballistic missiles) into asteroid killers.[25]

One problem with the explosion strategies is the risk of merely fracturing an asteroid in place, leaving more or less the same mass on the same collision course with Earth, instead of pushing it or dispersing the pieces. Another problem is that promoting nuclear explosives could increase risks from them here on Earth, maybe more than they would alleviate risks from asteroids.[26]

While getting rid of dangerous asteroids is plausible, the proposed technologies are still way too undeveloped to be confident about. The well-known astronomer Carl Sagan also feared the potential for it to be misused to send an asteroid *to* the Earth instead of away from it. According to the Lifeboat Foundation, "We share Sagan's concerns and believe more effort should be put into detecting asteroids than deflecting asteroids" (lifeboat.com/ex/asteroid.shield). We should try to know when and where an impact will be as far in advance as possible. With a lead time of 100 years or more, an impact zone could be gradually evacuated at a deliberate pace, even while uncertain methods were implemented to try to push or destroy the asteroid. Even a large city could be moved or dispersed within 100 years in a controlled manner. If Seoul had started moving 50 years ago to get out of range of North Korean artillery, the job would be half done.

On the other hand, if an impact warning became available only days or weeks in advance, emergency evacuation would be needed. Some cities have such plans already. Houston's hurricane evacuation plan, executed for Hurricane Rita (poorly) in 2005, involved converting inbound lanes on major highways into outbound lanes, thereby doubling the amount of outbound roadway. All areas should be required to have and test plans.

Since most of the Earth is covered with deep water, most asteroid impacts will occur in deep water. Like earthquakes, these can cause tsunamis. A relatively small tsunami destroyed the Fukushima I Nuclear Power Plant in Japan in 2011, releasing large quantities of radioactive contaminants. A disastrous earthquake-linked 2004 Indian Ocean tsunami caused over 200,000 deaths. Legends of ancient, otherwise unrecorded tsunamis saved many Andaman Islanders from that disaster, because they recognized the signs of an incipient tsunami and ran for cover. Flood stories of many cultures, including our own (think Noah's Ark), testify to the importance of maintaining a worldwide tsunami warning system as well as emergency evacuation plans for coastal communities large and small.

Finally, a couple of recommendations that are — literally — less earthshaking. One is for the reader to check out some shooting stars. They are space rocks, too small to cause damage, that burn up high in the atmosphere, putting on brief but awe-inspiring shows. Shooting stars can and often do happen on any night, but there are more during meteor showers. The Perseid shower peaks around August 12–13, typically at a rate of dozens per hour. Meteor "storms" are meteor showers with particularly high rates, hundreds or at times even thousands per hour. The so-called "king of meteor showers" is the Leonid shower, peaking yearly on a night on or near November 17–18, with peak rates that vary greatly over a 33-year cycle (because comet Tempel-Tuttle is the source of the Leonids and its orbital period is 33 years). One can get an idea of how many shooting stars will be visible on what nights of what years and from what meteor showers at leonid.arc.nasa.gov/estimator.html. Is there too much light pollution, it's cloudy, or you just don't feel like going outside? Then view videos of shooting stars available online at sites like YouTube.

Another thing you can do is hunt for micrometeorites. As discussed earlier, anyone can do it!

A final suggestion: Check out an old impact site. This can be part of a fun vacation. For example, Meteor Crater in Arizona has been commercially developed into a tourist attraction and is easily accessible from Interstate Highway 40. If you have more time and money, you can arrange a Tunguska site tour.[27] Such activities can definitely impress one with the awe-inspiring power of the universe.

Chapter Twenty-Seven

Sic Transit Humanitas: The Transcent of Man

Ten million years ago our ancestors were like us in many ways, but were not human. Ten million years from now our descendants will probably not be human. Were the Neanderthals human? Genetic studies show that their blood runs in the veins of modern humans. The past gives hints about future possibilities.

When was the last time you went through the old family photographs? Imagine doing it again now, but in an unusual way. This photo collection is titled "Time Machine." The first picture is of you. Next to you is a picture of one of your parents, and next to that your parent's parent (your grandparent), then your great-grandparent, and so on — back through time.

Four generations would take you back roughly a hundred years. Forty — just a few pages of an old-fashioned photo album or a couple screens of thumbnails — would go back a thousand years. That may seem like a long time ago, but we're not finished. Four hundred generations would be about ten thousand years. That is certainly a long time, yet the photos would still fit in an old-fashioned photo album, or alternatively perhaps 20 screens of thumbnails. Yet ten thousand years ago things were very different: There was no reading or writing (or school), no religion we would recognize today, and many other differences. The agricultural revolution was underway, and maybe you can fill in some of the other blanks. Yet, the most distant of your ancestors pictured, from some ten thousand years ago, would look similar to the most recent picture — of you.

A few interesting differences (besides the clothes) would, however, likely be present. If you happen to have an Andean background, you probably have a gene for high altitude-adapted blood that this ancestor did not have, because genetic adaptation to high altitude spread more recently.[3] On the other hand, if your background is Tibetan, you probably have a different and much older gene for that purpose. If you have a Scandinavian or East African dairy farming background, you probably have a gene for dairy digestion that this ancestor did not.[4] Europeans in general were darker skinned than today. This was true until the skin lightening A111T mutation of the SLC24A5 gene spread throughout the population.[5] This process was incomplete (and possibly had not even begun) as recently as 7,000 years ago, the age of Mr. La Braña 1 (named posthumously by researchers), who was found in a cave in Spain preserved well enough for detailed genetic analysis. He was revealed to most likely have had relatively dark skin and blue eyes.[6] Thus, based on past experience, humans ten thousand years in the future will likely have some interesting differences from those of today.

The most obvious differences will be cultural. Present-day nations, nationalities, ethnic distinctions, and even religions will change greatly, perhaps beyond recognition. From a genetic standpoint, one dramatic possibility is that organized selective breeding may occur. Dogs have been bred for about fifteen thousand years, and it is amazing how much they can differ from each other as well as from their wolf forebears after such a short time. In humans there is little precedent for this process. But there is also little reason to doubt that deliberate selective breeding could potentially produce, in just a few generations, super-athletes, super-geniuses, super-conversationalists, super-sumo wrestlers (already done), and so on.[7]

Perhaps slightly less dramatically, a selective sweep could modify the human genome in short order. Consider long-term nonprogressors, the tiny percentage of people infected with the HIV virus who do not become ill from it and thus do not need medicine. They are called nonprogressors because they do not "progress" to obvious illness. Unless a very cheap and effective cure for HIV is found, over the next ten thousand years the genes that protect these long-term nonprogressors may become the rule rather than the exception.

Another apparently on-going selective sweep relates to alcohol. I recently witnessed a tall man exiting a liquor store on a Friday afternoon with a hefty payload — and the expression and demeanor of naked intent to get home fast and begin another weekend to remember — or forget. The past ten thousand years appear to have witnessed an ongoing selective sweep of the genetic ability to more rapidly metabolize alcohol, clearing it from the body with the enzyme alcohol dehydrogenase. Consequently, populations not exposed to much alcohol until more modern times, like Native Americans, tend to have higher rates of alcohol abuse.

The havoc alcohol causes abusers and their families is so great that it is nearly inconceivable that licensed sellers can sell it to abusers. Licenses should be required of purchasers. If each license was a plastic card with an embedded chip to record purchases and permit them only at non-abuse rates, privacy could be assured and a serious social problem alleviated.

Currently, the world is so awash in substance abuse that significant genetic selection for resistance to addiction may be occurring. Intriguingly, while some genes apply only to specific intoxicants, like the alcohol dehydrogenase gene, others may affect susceptibility to addiction in general. Why is this intriguing? Because the potential for addiction is built into the brain, more in some people than in others. That potential is likely there for a useful reason, with the addiction issue only an accidental and unfortunate side effect. If addiction is bred out of the human genome, brains, thus minds, thus society could change. But in what ways? The reader might enjoy speculating, because that is about all one can do right now, since the facts are not yet known. It would be good to know them, of course. Thus the scientific community should make investigation of this question a priority.

A hundred thousand years (4,000 generations) ago, things were different

You'd need about ten photo albums or a couple hundred screens of thumbnails to hold a snapshot of just one person in each of 4,000 generations. What would your ancestor from 4,000 generations ago look like? For one thing, he or she might be a Neanderthal. Stockier and with larger brains than ours, Neanderthals intermarried or at least interbred with regular humans, leading to a small but significant fraction of the current average person with non-African background having Neanderthal genes. Physical features of Neanderthals often considered "distinguishing" can actually be found in some people. For example, maybe you or someone you know has an occipital bun — a bump or protrusion on the back of the head. Other people have particularly heavy brow ridges, or thick, bowed thigh bones, or a barrel-shaped rib cage, and so on. The photo from a hundred thousand years ago in your collection might indeed be a Neanderthal, however, more likely it would not be. What would it look like, then?

It could possibly be a Denisovan. Less popularly known than Neanderthals, the Denisovans had very broad, robust fingers, based on the single finger bone found so far, in the Denisova Cave in Siberia. Remember the mention earlier in this chapter that Tibetans' gene for high altitude adaptation is "much older" than then thousand years? That is because it is thought to have been inherited from a child of a union with a Denisovan.[8]

One thing we can rule out as the subject of the 100,000 year old photo is a Flores Man (*Homo floresiensis*). This is the 3–4 foot tall species whose bones have been found on Flores Island, Indonesia.

Apparently more closely related to people than any other species at the time, these short, hobbit-like hominids[9] (if one hesitates to call them people, "animal" induces hesitation as well) survived at least until 12,000 years ago[10] and just maybe to the verge of modernity, possibly giving rise to the Floresian "Ebu Gogo" (grandmother who eats everything) legend. Conceivably Flores Man still exists, living unnoticed in jungles of the region.

If this picture from a hundred thousand years ago was not a Neanderthal, then what would it look like? Except for the Neanderthals, humans had not yet left Africa. Hence they had doubtless not yet developed light colored skin, an adaptation that enables the comparatively dim sunlight of higher latitudes to better penetrate the skin where it powers vitamin D production (at the price of increased susceptibility to sunburn and skin cancer). Similarly, today's regional variations in facial features had not yet developed, nor had any other present-day regional genetically based variations, both known (such as green eyes common among the Kalash of Pakistan), and unknown and waiting to be discovered in the future.

A hundred thousand years ago we looked a bit different, though still human (because we were human). Thus, it is likely that a hundred thousand years from now we will also look different — but still human. This will be about 4,000 generations, enough for rather modest selective pressures to cause significant evolutionary changes to the human species. Today's regional variations in appearance (facial features, stature, skin color), which did not exist a hundred thousand years ago, will be long since lost in the mists of history if civilization and its associated traveling and migration continue. If we can identify existing evolutionary pressures and extrapolate, thus envisioning their magnified effects, perhaps we could characterize humanity as it moves forward a hundred thousand years from now. If our brains are our essence, that essence will change in some ways.

But how? Science should identify as many ways as possible. Here is just one example.

Do you think a gene variant that confers a high ability to distinguish between fantasy and reality would tend to propagate to the next generation more successfully than a variant resulting in an average or below average such ability? I am not so sure, but suppose for a moment it would. It turns out that the paracingulate sulcus (PCS), a brain structure, correlates with and likely confers just such an ability.[11] Some people have two well-developed PCSs (one in the left hemisphere and one in the right). Some have just one. Some have none at all. Hence PCSs can be well-developed, missing, or anything in between. It might not be long before an eye witness in a high-stakes criminal trial will have the size and number of her paracingulate sulci splashed across news screens worldwide since, after all, being good at separating fantasy from reality is potentially relevant to eye witness testimony. Dating services will soon have to decide: Are people with similar PCSs more compatible, or is this a case where opposites attract? Perhaps if two individuals have the same degree of schizophrenia-induced hallucinations, the one with better PCSs would be more able to function. It is unknown if John Nash, the Nobel prize-winning economist and schizophrenia sufferer, has good PCSs or not. The motion picture about Nash's life, "*A Beautiful Mind*," is a fascinating account but does not discuss the PCS issue. (Einstein's brain, however, is in storage and could actually be checked!) Chimpanzees, interestingly, do not have paracingulate sulci.[12]

If good paracingulate sulci significantly benefit reproduction, in a hundred thousand years most everyone will have good ones. If the opposite, hardly anyone will! If they confer reproductive fitness in some circumstances but not in others, their prevalence may be much the same as now — leading one to wonder what the deciding circumstances are. Surely it would be useful for science to investigate this further.

A million years ago

Despite the foregoing, a hundred thousand years ago people looked like — and were — people. But go back a million years and things were different. For one thing, the current dearth of other species very similar to us (that is, other members of the genus *Homo*) did not hold. There were others, very similar in many ways, yet odd and different as well. Not only *Homo floresiensis* (the "hobbits"), but also *Homo antecessor* and *Homo heidelbergensis*, for example.

Antecessor lived up to about 800,000 years ago and was roughly the height of modern humans, though of more robust build. Bones showing damage from stone tools suggest that *antecessor* practiced cannibalism and used tools, sometimes at the same time.[13] Their brain size was about 20% below ours, which still counts as large and, presumably, intelligent. With a low forehead and not much chin, they are only known to have lived in Europe. *Homo heidelbergensis* lived more recently than *antecessor* and may in fact be evolved from *antecessor* (just as we will evolve into something different if we survive long enough). First discovered in the form of a jaw found near Heidelberg in 1907, *heidelbergensis* stood tall, about 6 feet (2 m) in Europe and often exceeding 7 in South Africa, and was muscular and strong. The European population of *heidelbergensis* may have evolved into the Neanderthals though we don't know for sure. The African population may have evolved into modern humans, which would make *heidelbergensis* our ancestral species. Compared to *heidelbergensis*, however, we have higher foreheads, flatter faces, and are smaller boned. (Biologically speaking, *heidelbergensis* was "robust" while we are "gracile.") In this we are not unlike a *heidelbergensis* child, illustrating our neoteny, meaning slowed process of development.

One frequent biological effect of neoteny is the tendency to retain childhood characteristics into adulthood, since if development is slowed it may never get as far as it once did. Neoteny is considered a broad characteristic of modern humans compared to our ancestors. That helps explain why our knuckles don't drag on the ground. Instead, we have relatively long trunks, short limbs, small brow ridges, small noses, high foreheads, and flat faces, traits more characteristics of juveniles than adults in other primate species. This trend may continue into the future. You may like to imagine what our descendants might look like if this happens (here is the answer, written backwards so you don't accidentally read it too soon: .noitinifed yb ynetoen fo ecnesse eht si hcihw ,ekil-dlihC).

To further illustrate the amount of evolutionary change that can occur in a million years, consider the chimpanzee (*Pan troglodyte*) and the bonobo (*Pan paniscus*), our closest living relatives. They split from a common ancestor species roughly a million years ago and thus their differences form an interesting animal analogy to the amount of change that might happen to us over that time scale. They look fairly different and act even more differently. The bonobo is a bit smaller than the chimpanzee, but behavioral differences are more dramatic. Far less aggressive than the chimp, the bonobo is known for its unique sexual life. While chimps are promiscuous in a sense a human can comprehend, bonobos raise to a whole new level the integration of sex into multiple facets of daily life. Perhaps there are human swingers and adult industry careerists who could empathize with bonobo society to a degree, but to most of us it's pretty alien. Bonobo philosophy raises the "make love not war" concept to a whole new level. Bonobo sex, both *hetero-* and *homo-*, has an important place in the everyday functioning of bonobo society, for example in smoothing over conflicts to avoid fighting. This seems to be absent in the larger, more aggressive and dangerous chimpanzee. Additionally, bonobo society is mostly matriarchal (female dominated) while chimpanzees are highly patriarchal (male dominated), with a dominance hierarchy in a troop placing essentially all adult females below all adult males.

Given the huge behavioral differences between chimpanzees and bonobos, it is likely that *Homo heidelbergensis* temperament, behavior, and society differed considerably from our own (as well as from traditional human tribal societies, which also often differ greatly from each other). Such behavioral differences may have exceeded the significant differences in appearance and physique. We will never know the behavioral details for sure, of course, but we can make some guesses.

The general developmental principle that humans are neotenized relative to their predecessors most likely also significantly influences our emotional development, resulting in human adults having some temperamental characteristics more typical of *heidelbergensis* youths than *heidelbergensis* adults. If one has trouble — as I do — believing that neotenized physical characteristics have any great value for physically dealing with the world (running around, finding food, etc.), then one must suspect that it is neotenous neural and temperamental characteristics that drove the neotenization of our species, with neotenous physical characteristics merely an accidental side effect. So what might those neotenous brain and mind-related traits be? One trait relates to the fact that retaining juvenile characteristics tends to delay adulthood. This means a longer childhood, which today's developed nations use for extended schooling. A lengthened childhood period helps here: as the saying goes, you can't teach an old dog new tricks. In fact, human childhood is so long that most other animals die of old age in the time humans take just to grow up. By extending youth, humans likely have a longer period of high neural plasticity, supporting improved ability to learn over a longer period of time. Still, like old dogs, old humans seem to learn less quickly. In the future, if neoteny proceeds further, that may change.

Dogs serve as more than a source of sayings. They also provide a great example of neoteny themselves because dogs are neotenized gray wolves. Domesticated at least 15,000 years ago, they have held on to their position as "man's best friend" in large part because of neotenized aspects of their temperaments. Remember that standard gray wolves are not man's best friend. Gray wolves would *eat* a man (woman, and especially child) if they figured they could get away with it. Gray wolves (fighting weight may exceed 120 pounds or 55 kg) are definitely not good with small kids, unlike so many dogs. Wolf puppies, however, are cute, fluffy, playful, and fun, like many adult dogs. (In fact, so are the fiercest lion and grizzly bear cubs.) Thus neotenized mammal adults like us (and dogs) may be expected to be comparatively friendly, sociable, and playful. Also cute: We appear genetically predisposed to find juvenile characteristics like short arms, roundish, flattish faces with short or no muzzle, and small noses endearing. Can you think of any politician that one might suspect benefits in popularity from possessing neotenously cherubic or "cute" physical features? Many other animals are analogously programmed.

The amounts and types of evolutionary differences between us and *H. heidelbergensis* suggest the amount and types of evolutionary change we ourselves may undergo in our next several hundred thousand years. Thus our brains may enlarge, perhaps by around 20%. Will that make every man an Einstein, every woman a Curie? Perhaps — and perhaps much more than even that. Einstein's brain was on the small side, but surely an extra 20% could be good for something. Moving from brains to bodies, with *heidelbergensis*'s robust physique and our gracile one, if this trend continues basketball has good long-term prospects, while sumo wrestling may eventually face some challenges. No need to panic though; at a time scale of hundreds of thousands of years there is plenty of time for sports franchises to adapt to changing times.

Getting back to neoteny, a continuation of the trend in that direction suggests a distant future of shorter limbs and longer trunks, baby-faced adults, and more playful, friendly cultures. Hopefully increased neoteny, which in a sense lengthens youth, will result in longer natural lifespans, since getting old is not exactly a sign of youth. This may help explain our relatively long lifespans compared to most animals.

In a million years our descendants may be surprisingly alien to present day human understanding. The only way to tell for sure, however, is to wait and see. Since most people do not expect to be able to wait that long, we simply need to use our imaginations.

Ten million years ago and more

It is thought that our ancestors of ten million years ago, roughly 400,000 generations, were hominids (Latin: Hominidae) who had not yet branched out into their current descendants — orangutans, gorillas, bonobos, chimpanzees, and us. The hominid family and our genus, *Homo*, are well-known, landmark taxa in our evolutionary tree. However there are also superfamily, subfamily, and other taxa that are instructive (or perhaps just confusing) to mention.

The hominid family forms a branch of the hominoid superfamily, which also includes gibbons. The hominids branch in turn into orangutans and the hominine subfamily (last syllable pronounced like the number "nine"). The hominines branch into gorillas and the hominin (no "e") tribe, which in turn branches into chimpanzees and bonobos, on the one hand, and the hominan (with an "a") subtribe on the other, of which we are the only surviving example. The hominans contain only one living genus, *Homo*, or humans, within which we, the *Homo sapiens* (meaning "intelligent humans") are the only extant species. This makes our genus quite unsuccessful compared to, say, the rodent genus, which contains about 1,500 species out of around 4,000 mammalian species in all. In any event this terminological mess of hominthises and hominthats may seem ridiculous! Maybe it is. However, one way to help remember the ordering is to keep in mind that the *homin-* terms are (almost) in alphabetical order from smallest to largest grouping: hominan, hominin, hominine, hominid, hominoid, with one exception. Hominid is out of order but at least next to the other 'd'-containing name, hominoid.

Some milestones of prehistory

About 85 million years ago,[14] primates split off from the rest of the evolutionary "tree of life" (technically called the Linnaean taxonomy, after Carl Linnaeus (1707–1778), the Swedish scientist who created this branching map of evolutionary relationships). Eighty-five million years is a fairly long time ago, even for biological evolution; our roots go deep. Apes diverged from monkeys about 32 million years ago.[15] Apes are called hominoids, the first and largest category named "hom-" out of several such categories. The great apes, a subgroup of apes technically called hominids, came along around 19 million years ago.[16] They include orangutans, gorillas, chimpanzees, bonobos, and humans. Moving into the farther reaches of the 1–10 million year ago time scale, a partial fossil of a likely common ancestor of the hominines (comprising the present-day species of modern humans, gorillas, chimpanzees, and bonobos, but excluding orangutans since they branched off earlier), was found near Nakali, Kenya. This ancient species is thus called *Nakalipithecus.*[17] This hominine is just under 10 million years old.

Gorillas split off from our common ancestors about 7 million years ago.[18] The progenitor of chimps and bonobos split off 5–7 million years ago.[19] That progenitor species enjoyed considerable success over time, spawning almost two dozen separate identifiable human-like species, which group under the category "hominin" (no "e").[20] Most are somewhat obscure to the average person. Among the better known are in the genus *Homo*, literally meaning "human," and include *Homo habilis* ("habilis" is Latin for "handy man" since they used tools) as well as *Homo ergaster* ("working man"). Researchers released the discovery of *Homo naledi* in Sept., 2015, the first species known to have cultural rituals because they placed the bodies of some deceased individuals deep in the Dinaledi Chamber of the Rising Star Cave, South Africa. Although only one species survives today, *Homo sapiens* (us), this remarkable species has pretty much achieved world domination. The jury is still out on our fitness to rule but that question will be resolved sooner or later. Perhaps you are reluctant to call such extinct species as *Homo habilis* and *Homo ergaster* human. If so, recall the well-known hominid Groucho Marx, who famously groused, "I don't want to belong to any club that will accept me as a member."

The dramatic changes over the past 10 million years may well presage a like degree of change over the next 10 million. But what sorts of changes? There are many possibilities, but let us focus on brain size and intelligence.

We humans are big-headed, and we're proud of that. Our brains are bigger in fact than all but a few very large animals, elephants and whales in particular. But we are way ahead of even those animals on other crucial brain measurements. Since the brain mass of a species tends to increase with species body size, but not as fast, a measure called encephalization quotient is often used to express the deviation of brain size from what would be expected for a given body size if the organism had an average encephalization quotient of 1.0. On that metric, human brains are over 7× bigger than expected for an organism of our size. No other organism is that high. One survey puts elephants and whales at 1.3× and 1.8× respectively.[21] Bottlenose dolphins get to a little over 5×, and their brains are in fact close in size to ours, possibly making them geniuses of the animal kingdom.

White-fronted capuchin monkeys get to almost 5× although absolute brain size is a lot less than for humans because their body size is so much smaller. By contrast, cats are 1×, while dogs are 1.2×. The humble opossum, at a mere 0.2×, might thus be characterized as better endowed with beauty than brains.

Even ignoring body size entirely, we are still way ahead on the number of neurons in our brains. Those large animals with brains bigger than ours actually have fewer neurons, which are the information processing units of the brain. By analogy, ordinary computers are getting more powerful year by year not because they are getting bigger in size but because their processors are being made with more information processing units. For computers these are not neurons but transistors. Transistors are actually getting smaller over time, which is why personal computing devices, despite their high computing power, are actually getting physically smaller instead of bigger (an example is the transition in personal computing from laptops to small tablets and smartphones).

A dramatic process of brain enlargement in our past began approximately two million years ago.[22] Our brains have literally tripled in size and, perhaps, number of neurons since that time. That is why chimpanzees like bananas, while some of us like not only bananas but (e-)books too. What if this process continued? Will our brains triple again over the next couple of million years, giving us descendants who chuckle condescendingly at our admiration for books as we might chuckle over a monkey's admiration for bananas? Would such brainy descendants consider our most vexing problems of war and peace, poverty and excess, illness and health, love and hate to have obvious solutions easily taught in elementary school? Perhaps they will, but there are serious limits to such rapid brain growth. These limits appear to forbid brain size (measured as number of neurons) from increasing at a rapid clip forever. In particular another tripling would, it appears, require major changes in brain structure surpassing the structural differences between human brains and those of other apes — in short, a major evolutionary leap requiring significantly more than a measly few million years. On the other hand, doubling brain size can be done more easily (and presumably quickly) because no major architectural changes would be necessary. In other words our brains can fairly easily double, but tripling would be more problematic. "Why?" you might ask.

Here is a mathematical explanation. First, the brain is composed of many parts that must communicate with each other. In contrast, the old syncytium theory that the brain is essentially a big blob of weakly organized tissue, sort of like a big ball of cotton, had been largely disproved by Spanish neuroscientist Ramon y Cajal as early as 1900, a feat for which he won the Nobel prize in 1906. Second, the bigger the brain, the more parts there are. The more communicating parts, the more neurons need to be devoted to communication (thus acting like the telephone wires and internet cables of the brain). This is a problem. Let's see why.

Suppose 3 parts of the brain, 3 people, 3 computers, 3 offices, or 3 of any kind of communicating entity all need to communicate with each other. How many communication paths are needed? Three: One between units A and B, one between units B and C, and one between units A and C. Suppose we add a 4th communicator, D, that needs to communicate with all the others. How many new paths are needed for all to communicate? Paths between D & A, D & B, and D & C. The number of these new paths, all involving D, is three, or four minus one: Unit D needs connections to the 3 others already present. We added one more communicating chunk and needed to double the communication paths. The problem just gets worse the more units we add. The thousandth unit requires adding 999 new communication paths running between the new unit and each of the previous 999. Thus while the fourth unit needed 3 more paths the thousandth needed 999! So adding one new section to a large brain that already has lots of sections requires lots of extra neural tissue to be added for communication purposes. In practice, this is the white matter of the human brain cortex, which because of this problem, forms a disproportionate fraction of the human brain compared to other primates. It gets progressively more biologically expensive to incrementally increase brain capacity and humans are in the zone where this expense significantly impedes dramatic increases.

Of course, doubling our brains is nothing to sniff at and could lead to impressively brilliant descendants, even if much more than doubling proves unachievable. These descendants would have big heads and probably wide hips, so that they can be born. We can only hope that such big-headed, wide-hipped, high-flying beings will still smell the flowers and, indeed, enjoy an occasional banana.

Recommendations

The future of the human race, to a significant extent, will be written in our genes. Much about us at present is written there now. But we know too little about what those genes do, what variations in them mean, and what beneficial new variations are possible. Such information will enable, for example, advanced medical interventions. The beginnings of this are occasionally in the news now. To add to knowledge of our genes it will be helpful to develop animals with some of their genes replaced by homologous human genes, so that these genes can be more effectively studied. For example "humanized" mice are now available and increasingly used in cancer research and other laboratories. Humanizing other animals as well would lead to still more knowledge. Applying this concept to brain-related genes, unusually intelligent dogs could come to replace ordinary pooches as "man's best friend."

We could advance our understanding of genes related to such human capacities as language and intelligence by studying the homologous genes of our nearest animal relatives. Take for example the FOXP2 gene, linked to our language ability though still somewhat mysteriously. The function of the FOXP2 gene has been investigated using mice by substituting the human version for the original mouse version and observing how this affects the mice.[23] It would be even more interesting to create a transgenic chimpanzee or bonobo (or perhaps a vocally sophisticated animal like the porpoise or prairie dog) that has its normal FOXP2 gene substituted by the human one. What would be the effects on the animal? This should really help to elucidate its function in humans. The same approach could be applied to other genes whose expressions are regulated by FOXP2, which works by regulating other genes — it is a "transcription factor" gene. One or more of these other genes may be even more closely connected to language ability. Any other gene linked to the unique aspects of human brain structure and function could also be investigated with this approach.

Many important conclusions remain to be discovered by comparing our genome with those of related animals. The animals most closely related to us are the primates. Yet many primates are decreasing in population and are now, or may become, at risk of extinction. Every such extinction will close off access to genomes and associated phenotypes (traits) that have much to tell us about our past and, maybe, our future. Thus, preservation of primate species is in our interest, hence recommended. The diversity of gorilla populations is one example of concern, as numbers of gorillas in distinct population groups are decreasing precipitously. Much is at stake: Understanding how we got where we are now can shed considerable light on where we could go, what our genetic potentials are, and how long it might take to reach them.

Human evolution has produced great change, and great strides, over the past several million years. But wouldn't it be nice if changes we might desire — much smarter brains, markedly more athletic bodies, adaptations helpful in colonizing other planets and moons, resistance to diseases from malaria to flu to heart disease, much longer lifetimes, ability to reproduce without need of assistance from the opposite sex (most plausible for women), inborn dislike of the taste of junk food or, alternatively, ability to nutritionally thrive on junk food (since it is soooo tasty!), three hands since everyone knows that sometimes two are not enough.... It would be even nicer to get such things without those annoying multi-million year lead times! Certainly in ten million years "we" will look very different regardless, but changes can potentially happen vastly faster if we manage and control genetic change appropriately. Far from the old eugenics movement of the decades surrounding the year 1900, which was so scientifically naïve and blatantly racist as to make one doubt the mental fitness of its proponents, a new movement would be aimed at encouraging genetically unusual people to be created and to exist, instead of the discredited and evolutionarily regressive concept of artificially discouraging out groups from reproducing.

How might this new "benegenics" approach work? Benegenics would involve first, screening people for new and rare mutations. It is those genes that will eventually rule the future. It is also those genes with the greatest potential to help change the human condition — for the better, but maybe for the worse if we are not careful and wise. While by definition a very small proportion of people have genes that are rare, the total number of such persons is larger now than throughout all of human history and prehistory. The simple reason is that there are more people in existence now than ever before.

Once identified, such people may often be willing to participate as research subjects to expand human knowledge about our genetic potentials, particularly if paid. As for the rest of us, we are each unique in our combination of genes, but that uniqueness is not passed on to our descendants. Our children contain their own unique genetic mix, but that mix is composed of the same genes found in countless others. Your genetic recipe is unique but the gene ingredients are standard. The ingredients get passed on down the generations but the recipe gets changed each time so much that, though someone might well look like their parent, any resemblance to a grandparent would be much less and, beyond that, hardly at all. The contribution one may hope to make with one's children is not genetic but social and cultural, because constructive people benefit the world and every little bit helps.

Even genetic differences among ethnic and racial groups are minor contributors to the human genetic range, because genetic diversity within ethnic and racial groups is known to dwarf average differences across groups. The few genes that cause visible distinctions between some groups may seem noticeable but are often only skin deep (literally) and do not determine deeper aspects of the human condition. Consequently the self-serving pipe dream of the old eugenicists, that suppressing reproduction of people unlike themselves would improve the human race, is naïve. It is also socially destructive, thus contradicting the movement's own stated goal of a better society. In contrast, it is those rare individuals who really are different — with new gene variants for better brains and bodies, who will contribute genetically to the distant future.

The Seventh Generation:

The Next Hundred Million Years

Chapter Twenty-Eight

Floating Prairies of the Seas

The surfaces of the oceans are surprisingly empty. There is an ecological niche there and it could get filled, whether by nature or by design. Things will get interesting and different, if and when it happens.

The natural prairies of the North American Great Plains, now mostly gone, are like rainforests in important ways but a lot closer to the ground. They are both lush, dense, complex layers of life covering the land surface, bursting with vitality. But this concept is not implemented on the ocean surface. But perhaps it could and, in the fullness of time, will be. It could be done by genetic engineering much sooner, probably in our own time, if desired. By thus increasing the biological productivity of the vast ocean waters, which cover 3/4 of the Earth's surface, the extra photosynthesis could remove a lot of carbon dioxide from the atmosphere, helping to control global warming.

The ocean surface is vast, with plentiful water and sunshine to nurture plants and animals from microscopic to blue whale-sized. Surface waters often contain plankton, small organisms that drift with the currents. Algae, for example, use sunlight and dissolved nutrients and gases to grow, and the many tiny animals among the plankton eat the algae and each other. If you are ever unfortunate enough to be a castaway, adrift at sea, a survival tip is to use a nylon stocking to sift plankton out of the water. This will often capture nutritious food. The largest creatures to ever inhabit the Earth, blue whales, eat only plankton. More specifically, those enormous animals eat mostly krill, small shrimp-like arthropods that form part of the plankton community.

Although plankton live *near* the surface, they don't live *on* the surface. The ocean surface itself is generally pretty clear. Looking down at the water, one usually sees mostly water, not a plant covering. Thus, the competition for sunlight that often seems to characterize plant life on land seems not to be as important in the oceans. This is due to dissolved nutrients being present only in dilute form, as well as the action of currents, waves, and wind. On land, plants that shoot up the fastest and tallest get more precious sunlight, solar energy that they use for growth. They then cast shade on neighboring, shorter plants, depriving them of energy and tending to impoverish them. That seems to be why trees are tall and why rainforests grow as high as they do. Prairies display similar traits, though on a shorter scale, probably because periodic prairie fires tend to kill off tall vegetation like trees that rely on expensive structural investments (like trunks) for which the height investment tends to pay off only after years of delay. "Grow fast, die young" seems to be the operative strategy on the highly competitive prairie. Less clear, however, is why the race to grow higher than the competition has not taken over the seas. Maybe ocean vegetative life just needs more time to pursue that evolutionary strategy and someday will achieve it. The fresh water environment already has succeeded.

Fresh water

The value of taking over the water surface is sometimes seen in fresh water ecosystems. Fields of duckweed, the world's smallest flowering plant, sometimes coat the surface of stagnant ponds. Each plant's tiny rosette of leaves floats on the water, with small roots descending downward. They can proliferate to the point where the water appears to be covered with a green sheet.

Even more dramatic is the common water hyacinth (*Eichhornia crassipes*). Not content to merely float at surface level, it can rise into the air, covering the water in such profusion that it looks like a solid meadow. This can happen surprisingly fast, as populations can double in as little as two weeks.

Though beautiful (Figure 12), water hyacinth can be a viciously invasive pest outside of its natural habitat. Although so dangerous it is illegal to transport in many areas, it can purify polluted water and, when cooked, the young leaves, stems, and flotation bladders are edible.[1] The bladders may be deep fried and are more nutritious than French fries. (Caution: Do not eat if grown in polluted water. Check local regulations for legal restrictions on transportation before transporting to your dinner table.[2])

Figure 12. Water hyacinths (source:
US Dept. of Agriculture).[3]

If the common water hyacinth grew as profusely in salt water as in fresh, the world's oceans could be radically transformed. But they only grow in fresh water.

The oceans are a different story

Mostly, below-surface plankton rule the waves. Floating mats of algae occur sometimes. The Sargasso Sea (in the storied Bermuda Triangle) has historically been known as a repository of masses of floating seaweed, though its density is far below the level of old seafarers' legends.

If floating seaweeds are destined to cover the ocean surface, and thereby out-compete plankton and underwater seaweed in the struggle for sunlight, they will first have to solve some problems that, so far, they clearly have not. The three biggies are mobility, nutrition, and predation. The mobility problem occurs because, left to be buffeted by wind, wave, and current, a worldwide carpet of floating seaweeds would soon give way to giant swaths of cleared water as massed seaweeds pile up elsewhere to die. The nutrition problem derives from scarcities of life-supporting substances dissolved in the water (plants cannot live by water alone). The predation problem arises because, by the inescapable nature of life as we know it, plants are of interest to hungry animals. However, appropriate changes in the plants' genomes could potentially produce solutions to all of these problems.

The mobility problem

To have a hope of covering any ocean surfaces, floating seaweeds will need to confront the fact that anything floating on the ocean surface is subject to motion from currents and winds. Such motion can easily lead to landfall and death of the organism *en masse*. This actually happens occasionally to a strange little blue, floating, vaguely jellyfish-like animal called the "by the wind sailor."[4] They have little non-movable sails fixed so they sail at about 45° from directly downwind. This keeps them safely in the wide open ocean enough of the time for them to breed and maintain a permanent population. Floating seaweeds could be genetically engineered to behave similarly.

Floating seaweeds which do not have sails have a better opportunity to maintain stable off-shore populations in locations that either remain stagnant, unaffected by currents and winds, or circulate in a repeating loop (these large and slowly whirling ocean areas are called gyres). The most obvious qualifying body of water is the Sargasso Sea. Other areas are known to scientists as the Great Pacific

Garbage Patch, the North Atlantic Garbage Patch, and the Indian Ocean Garbage Patch. These are oceanic areas where small particles of plastic accumulate in the water where they are deposited by surrounding currents. If plastic trash can build up, the water is stable enough for seaweed to accumulate as well. Thus floating prairies of seaweeds or other vegetation (such as salt-resistant water hyacinths to be created by genetic engineering, or eventually evolving naturally over the vastness of time) could develop in those areas, reaching out of the water to form apparently solid meadows that hide the water beneath. Colonization by the smaller land creatures would follow naturally, as they could climb among the prairies without falling through and drowning. For such prairies to exist, however, requires solving another problem, nutrition.

The nutrition problem

The main things vegetation needs — water, oxygen, and carbon dioxide — are readily available from the ocean and atmosphere. Nitrogen is critical as well, is present as 78% of the air, and can potentially be extracted from the air and chemically transformed into biologically usable form by advanced biochemical pathways. Legumes (such as beans) do that on land already with the assistance of special bacteria living in nodes on their roots. This helps to account for the success of the legume family. This proves it can be done and therefore that seaweed could potentially do it someday.

Smaller quantities of various minerals are also important. For example iron, an essential trace mineral for plants, is in short supply in many parts of the ocean and this limits phytoplankton growth. Indeed, in a 2012 report, iron compounds were dispersed in ocean water, stimulating growth of masses of diatom-rich plankton.[5] These organisms subsequently died and sank, bringing down large amounts of carbon from carbon dioxide gas they had absorbed, presumably to the ocean bottom to be sequestered out of harm's way for a long, long time to come. Since diatoms, ordinary plants, and other photosynthetic organisms take in carbon dioxide (and output oxygen) this is a sensible strategy for controlling atmospheric carbon dioxide levels.

In fact, every naturally occurring element is present in seawater even if the concentration is too low, as in the case of iron in many areas. More gold than King Midas ever dreamed of is dissolved in every cubic mile (3 km^3) of seawater. Not that vegetation needs gold, but everything it does need is there to be extracted from the water. The problem is that growth is limited by the low concentrations of key substances. Vegetation will need to solve this if it is to become densely packed on the ocean surface, but it can in principle be done.

Reduced to the basics, the problem is that modern floating seaweeds can access dissolved nutrients only very close to the surface, because that is where the seaweed is. But if they could get to any nutrients present in the water 10× deeper than they can now, they could get 10× as much nutrients. What is needed are long, thin root filaments that descend into lower waters and take up the nutrients down there, transporting them to the floating seaweed at the surface. Such roots may evolve of their own accord, given enough millions of years. Alternatively, they will be genetically engineerable once the genetic coding language is deciphered and genetic engineering becomes a branch of software engineering.

What would it take, biologically, to grow such roots? To get an estimate, suppose a root filament has a dry weight roughly the same as a human hair, about 0.05 milligrams per cm.[6] Its thickness in water would be considerably greater than the thickness of a hair, because to be alive the root filament would need to contain water as a large part of its volume. At 0.05 mg/cm of dry biomass, that is about half a gram for a root filament the length of a football field. Descending downward into the ocean, it would contact a lot of water from which nutrients could be extracted and transported upward to the rest of the seaweed. For a significantly sized floating seaweed clump, only half a gram of biomass (less than a fiftieth of an ounce) is not that expensive to grow and may be well worth the effort considering the benefits even one such long root fiber could have for taking in nutrients. Indeed, a root shorter than a football field would be even cheaper to grow and probably still long enough.

Predation

Fish and other sea creatures like to eat underwater seaweed and algae. Land animals like to eat land plants, but that obviously doesn't prevent thick profusions of plants in prairies, rainforests, and other land habitats. Protective mechanisms are often used by land plants, such as poisons and prickers. If oceanic prairies come about as a product of genetic engineering, they could perhaps occur in decades. If on the other hand these prairies of the seas develop naturally, humans may be long gone by then though we may hope otherwise.

Will Earth's oceans someday contain vast expanses of seemingly solid floating islands made of floating green masses sprouting from the water's surface? Only time will tell.

Recommendations

Genetic engineering could be used to create oceanic prairies that are useful to humans. The plants should be designed to be as inedible to other animals as possible, or even to be able to eat some of the animals that try to eat them, perhaps like a Venus flytrap eats insects on land. Until this occurs, culinary use of water hyacinth should be promoted as a valuable menu addition.

Chapter Twenty-Nine

Get Ready for the Greenish Revolution

Plants may be changed to be more useful in many strange ways. But why stop with making plants better? Why not make animals whose skins can photosynthesize, thus needing less food? There are animals that already do that, and genetic engineering could help take this concept to its strange but logical conclusions.

Insufficient food is a fundamental threat. It is the limited capacity of the Earth to produce food that, more than anything else, imposes a limit on human population. This finite capacity for food production conspires with the Malthusian tendency of populations to expand to the limits imposed by available resources. Thus unless human population increase ends across the board as a result of contraceptives, urbanization, and other voluntary checks, too many people through no fault of their own will eventually find themselves in a state of chronic food insecurity.[1]

When many people are preoccupied with just feeding themselves and their families, those people cannot efficiently develop their potentials, inhibiting advances in the world's science, engineering, religion, entertainment, economic, and other societal activities. Thus if food security could be dramatically improved, humankind in general would benefit greatly.

Agricultural productivity through mechanization

In wealthy and technologically advanced countries, for example the US, UK, and Australia, agricultural employment is in a long-term decline. In the UK, agricultural employment declined from 700,000 in

1984 to just under 550,000 in 2004;[2] in the US the percentage of the labor force devoted to agriculture went from 33% in 1900 to a mere 2% in 2000;[3] and in Australia the percentage decreased from 6.3% in 1986 to 4% in 2004.[4] This is certainly not causing food production to become inadequate — obesity is a problem in all three countries.[5] Rather, efficiency, in the form of food production per person-hour of agricultural labor, has been increasing. Advances in automation have enabled this.

Such technologies have reached an impressive level. Standard center pivot irrigation systems can water plants while rotating over circular patches of crop-laden land literally a full mile (almost 2 km) in diameter. You can often see these patches from airplane windows when flying across the US. Prices of these machines in dollars will set you back six digits. That might sound like a bargain for a precision machine that big, but at those prices they won't be purchased to increase farming productivity in third world countries any time soon. The situation is analogous for large farm harvesting combines. These vehicles are essentially grain processing factories on wheels, often with GPS tracking and numerous other high tech components.

Improving labor efficiency of food production is a happy trend, so we should make sure it continues. Yet, at least in the developed world the potential for improvement is limited simply because food production is so labor-efficient already. If only 2% of the workforce does agriculture, reducing that to 1% would require doubling agricultural labor productivity, but that would make little difference in the scheme of things. Instead, we need to look at other methods for boosting agricultural productivity besides the expensive methods (like mechanization, chemical fertilizers and pesticides, etc.) that have been so effective in developed countries. Let's explore this next.

Secret of the bison?

The plains bison, a subspecies of American bison, formerly ranged the great prairies of North America. Feeding on naturally growing vegetation, bison of course did not domesticate their food plants nor did they ever employ farming techniques. They mostly just wandered around eating. Often they did not flee from hunters, nearly going extinct as a result. How did such clueless beasts historically enjoy such a plentiful food supply? It turns out that natural ecosystems (prairies, rainforests, etc.) have high ecological productivity, which we might measure in terms of the quantity of plant and animal tissue produced per unit of time.[6] In fact, the greater the diversity of species in a natural ecosystem, the more productive it tends to be.[7] Unfortunately, unlike bison, people can't eat random plants growing in a prairie.

If humans found wild plant tissue edible, it would be good for the environment. We merely need to figure out how to make it happen. Less (as in no) pesticides would be needed. Natural habitat would be less disrupted, since it could be used sustainably rather than cut down, destroyed, and replaced. And with Mother Nature as the farmer, less labor would be required, making labor efficiency high. The big problem is solving the edibility issue. As it stands currently, productivity of natural ecosystems is a lousy approach to getting dinner on the table.

An ideal solution is the "analog food mill" envisioned by Macfarlane in 1967.[8] You stuff any plant or animal material into the input hopper and it extrudes a "square strip of heavy paste" which can be flavored to taste like chicken, fruit, or cheese — not to mention govond and oegel, whatever they are. It did this while "chuffing" out quantities of inedible dry waste. (Macfarlane had it also able to serve as a small bomb by throwing it against a rock or hard wall, as long as the user was not planning on getting hungry later.) Presaging the future, in 2013 the principle was actually demonstrated in the lab, with a process for converting indigestible cellulose from arbitrary plants into healthful and nutritious amylose starch.[9] This could become a big deal. Not only would it greatly increase the food supply by making food from inedible plant material like corn leaves and husks, it would reduce the need for farming because wild plants on uncultivated land could be harvested and converted into food.

What we can do

Recall that one very labor-efficient form of agriculture would be an ecosystem in which things are permitted to grow by themselves, essentially without interference, but in which all the growing plants (and animals) are edible. This futuristic concept would require a radical transformation in existing ecosystems emphasizing composing them of edible organisms. In principle, given some specific geographical area, it could be made to work there. Since high species diversity tends to increase ecosystem productivity, it will ideally contain many different edible species. Variety is a key trait of a healthy diet, so that is another plus of the edible ecosystem paradigm. However many of the different foods might be unusual, presenting a wide spectrum of taste characteristics that would take getting used to for many people.

To start to gain a practical understanding of the concept of new kinds of edibles, perhaps the easiest thing one can do — and one that is likely to leave a good taste in your mouth — is to wander the aisles of the nearest different ethnic grocery store. These stores usually have interesting and unfamiliar foodstuffs like canned exotic fruits (think breadfruit, etc.), many different seasonings, and all kinds of other interesting items one can try. Yum!

Another way to start to get a taste of the concept is to identify, and try, the edible wild plants that live in your area. This requires a bit more research since one must learn to identify, locate, and prepare such plants. Mistakes are not recommended because many plants have toxic properties.

If all the plants in the area are edible, the situation is much safer, not to mention tastier. To make sure they are edible, a technology that

287

seems realistic within the next several decades is small solar powered robots, programmed to recognize edible plants and to kill competing inedible plants. Simply turn loose a herd of these robots and it won't be long before general edibility is assured. Such robots would also be quite useful for weeding conventional farms and even edible home gardens, thus providing even more incentive to invent them.

Speaking of edible gardens, another interesting personal step one can take now is called edible landscaping.[10] This is simply the practice of using plants in your yard that are both edible and ornamental, instead of just ornamental. For example, many people plant trees. Fruit trees can be just as nice as other trees, while providing delicious, healthy, and inexpensive treats. Indeed, wherever plants are desired, edible ones can often be obtained that are just as visually appealing as other plants. Edible landscaping is a step in the direction of the edible ecology paradigm; why not try it yourself?

The genetic engineering approach

Another strategy, not discussed above, is genetic engineering. It could be applied to achieve the goal of making wild plant species more edible while maintaining or improving their ability to compete in their ecology. This would require considerable research as well as updates to the current US regime for gaining permission to grow ("deregulating") transgenic plants. The current deregulation process is so expensive that only major commodity crops are cost effective to deregulate. Yet genetic engineering has a whole other dimension of possibilities as well, as we see next.

Second secret of the bison — the human connection

Long term, the nutritional value of plants would be greatly increased across the board if cellulose, a major component of plants, could be digested. Bison can do it. So can horses, cows, and other ruminants. Termites can too. Their trick is to host symbiotic microorganisms in their digestive tracts that break down the cellulose into sugars, which are then easily absorbed. Humans host symbiotic microorganisms in the gut too, but they are not the kind needed to digest cellulose. This suggests an obvious target for genetic engineering: Why not genetically modify human intestinal bacteria to do for us what other intestinal flora do for goats, sheep, bison, cows, horses, and termites? Maybe eating grass and such would require some degree of intestinal fortitude. The Defense Advanced Research Projects Agency (DARPA) tried to make that work for soldiers with their Intestinal Fortitude project.[11] Success could help alleviate human hunger in civilians around the world as well. Thus it certainly seems to make sense to prioritize this.

Food from the Sun

Better ways to eat plants are good goals, but of the plants themselves, the vast majority don't have to eat (although a few, like the Venus flytrap, are carnivorous). Instead of food, plants get their energy from sunlight. Wouldn't it be useful if humans could too? Alas, we can't, at least not yet, but coral, sea slugs, and even giant clams can.[12] So do some large snails.[13] Another example is the flatworm species *Convoluta roscoffensis*, which sometimes appears densely on beaches in the UK and turns them green. Green, because of the green algae that live in their translucent bodies. They pick up this algae when they are born by eating the egg cases from which they emerge; certain algae have attached to the egg cases in anticipation. These algae take up residence inside the worm between cells of its body. As the worm matures, its digestive system degenerates, it can no longer eat, and it begins to rely for nutrition on the algae that live in it. The worm must then seek sunlight to support the algae. According to James Oschman, "Upon loss of the theca [sheath, covering], the alga assumes an irregularly shaped form. Fingerlike processes of the algal cells penetrate between adjacent animal cells."[14]

Although this may sound weird, it probably doesn't hurt and the worm probably doesn't mind.

Increasing the degree of plant-animal integration further, the method used by coral and various other marine animals is to have symbiotic algae living, not between their cells (like those flatworms), but actually inside some of their cells. The algae are typically of the genus *Symbiodinium*, and live in "symbiosomes," blobs inside the animal cells that hold the algae separate from the rest of the cell. Each symbiosome is a kind of really tiny terrarium (a "nanoterrarium") maintained by the finely engineered nanotechnology device of nature we call the cell. The cells supply the algae, in its symbiosome home, with basic chemicals and exposure to light. In return the algae produce nutrients that the animals extract from the symbiosome and use. In coral, when these algae die the coral loses color and, if not reversed, itself dies in the phenomenon called "coral bleaching."

The degree of integration can be tighter still. Observe that algae (like their descendants, the plants) do photosynthesis using chloroplasts. These are small green organelles, organelles being the tiny nanomachines that serve as "organs" of cells. Chloroplasts thus give plants their green color. The chloroplasts are thought to have once been independent organisms that, eons ago, took up residence inside cells of other organisms, where they have lived ever since.[15] What about animals whose cells can contain chloroplasts directly, eliminating the inefficiency of using algae as the middleman?

The Sacoglossa, a group of about 300 species of sea slug, are sometimes called "solar-powered sea slugs" because they steal chloroplasts from algae that they eat, incorporating them into their own cells.[16] However these chloroplasts do not last long, so to replenish their supply the sea slugs must eat more algae. What we need are animals that not only maintain chloroplasts in their skin cells indefinitely, but also pass them down to their offspring. Though far beyond our current capabilities to achieve in arbitrary animal species by genetic engineering, eventual success could boost the fitness of such species by reducing the amount of food its members would need. This could help endangered species to survive, make animal husbandry more efficient, and eventually put species without chloroplasts at a competitive disadvantage.

Up from the lower animals

People and other higher animals should photosynthesize, whether by using extracellular algae like the *convoluta*, intracellular algae like the giant clam, or chloroplasts like ordinary plants. There is no point in restricting this valuable food production technology to plants and lower animals like worms, clams, snails, and sea slugs. If humans could photosynthesize, we would develop an attractive, healthy-looking greenish sheen, and would be making some of our own food from sunlight, further advancing food production by having some of it done by our own bodies. The problem is we don't know how to make this happen. Microorganisms do live in and on us, in all healthy people as well as those individuals (like the famous case of "Typhoid Mary") who are long term carriers of a disease. But they don't photosynthesize. Technologically it would be a tough task, and achieving it would be a long-term goal involving significant genetic modification of suitable microorganisms, as well as potentially animal and even human genomes. This would clearly be more feasible to do with animals before people. But the benefits would be impressive, making it eventually worth considering for people as well.

Beyond the Seventh Generation:

Farthest Reaches of the Future

Chapter Thirty

Accelerating Evolution

The diversity of species has been increasing on Earth over the course of biological evolution. With another four billion years or so to go, the Earth could eventually become riotous with different and diverse life forms, far more than it has ever been in the past.

The transformation of a disordered collection of molecules into a unitary bird, with its strong, lightweight, complex feathered surface, on the order of 700× denser than the air it floats through,[1] is as amazing as the most exotic and enormous deep space cosmological phenomenon. From the smallest life form, microscopic nanobacteria about 200 nanometers across and first photographed in detail in 2015,[2] to the hugest, named Pando, a multi-stemmed quaking aspen "plant" over 100 acres in size, they are unlike distant stars and galaxies in that we have convenient access to them. Perhaps the degree of biological change that takes nature hundreds of millions of years will soon be possible to achieve by human design within the span of a single human life. Certainly, nature appears to be slowly getting better at biological engineering, because the number of species is increasing over time (see Figure 13).

The world's number of different kinds of organisms has been increasing, on average, for hundreds of millions of years — a long time.[3] Things did seem to plateau for a while, from around 450 million to around 250 million years ago. But since then, it's been back to normal, with the number of different species following a generally increasing trajectory punctuated by occasional downward plunges (Figure 13). These plunges are caused by dramatic mass extinctions, which according to one analysis approach have a curious tendency to occur roughly 62 million years apart.[4] Although the jury is still out on the causes of these mass extinctions, one interesting theory is that changes in atmospheric oxygen levels are a factor because they stress the biosphere. Genetic markers consistent with this in fact have a periodicity component of 61 million years.[5] A plausible scenario for this is that reductions in ocean currents cause the oceans to stagnate, leading to loss of oxygen in the water and thus its ability to sustain life.

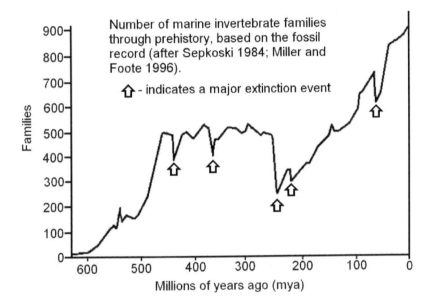

Figure 13.[6]

Based on the 62 million year cycle time, it appears we are due or even overdue for the mass extinction that is occurring now. This one is so recent that it does not yet show up in Figure 13, with its time scale of hundreds of millions of years. We know the cause of the current one, however. It is due to human-caused activities such as habitat destruction (e.g. cutting down rainforests) and climate change. Things would likely get considerably worse if ocean currents and other oxygen level regulating processes fail to remain stable through this. Unfortunately, human-caused global warming is thought by some climatologists to present a significant risk of just such an ocean current stagnation and the consequent oxygen loss. The underlying tendency posed by the 62 million year cycle, the current new human-caused climate warming trajectory, and the on-going wave of extinctions presents a bad luck trifecta.

The current situation disrupts an apparent very long-term megatrend of increasing species diversity. From the perspective of geological time scales of a hundred million years or more, human influence is decidedly short term. Long, long after we humans are either extinct or have changed into something more different than the difference between what we are today and gorillas, lemurs, or even squirrels, the Sun will still rise at dawn and set at dusk. Hopefully, life on Earth will continue on for a long, long time. This brings us back to the long-term overarching megatrend of increasing species diversity.

How new species come into existence

Evolutionary theory has long been uncertain about important aspects of how new species come to be. The process has no single name, perhaps owing to the subtleties of species creation: Speciation, divergence, branching, cladogenesis, splitting, radiation, and diversification are all similar terms for this process. This team of terms conceals a collection of causes and a mixture of mechanisms. Here are a few.

- Allopatry (meaning "other fatherland"): The range of a single species is split, such as by the birth of a mountain range or by rising waters, and the two populations evolve separately, becoming more dissimilar until they are separate species. This seems unlikely to happen to humans in a mobile global society, or to other species with near-global distribution, but has occurred many times with other organisms.

- Peripatry ("nearby fatherland"): A small group ventures beyond the periphery of its species' range and ceases to breed outside of its own group, evolving in a different direction until it is a separate species. This could happen to humans as a result of a successful self-sustaining extraterrestrial colonization project.

- Parapatry ("beside fatherland"): Within the continuous range of a species, members in different subranges breed preferentially within their own subrange, permitting the gene pool of each subrange to evolve differently from that of the other(s). The differences accumulate until, where before there was one species, now there are two (or more). This explains the various species of Darwin's famous finches on the Galapagos Islands.

- Heteropatry[7] ("different fatherland"): A single species occupies different ecological niches within the same geological range, leading to separate breeding populations and accumulation of genetic differences until the populations are different species. For example, an insect species that likes to eat two different kinds of fruit might, over time, split into two species, each preferring a different kind of fruit.

- Sympatry ("together fatherland"): A species divides into two even though both groups occupy the same geological area. It occurs most commonly through heteropatry and may be mediated genetically by processes like "reinforcement" and "chromosomal inversion."[8]

- Niche splitting: A species adapted to a niche that spans a continuous spectrum of conditions may produce offspring that are specialized to part of that spectrum at the cost of being less suited to the rest. The specialized offspring breed true, producing a new variety of the species that displaces the rest of the species from that part of the niche. The original variety, displaced from part of its original niche, specializes to the remaining portion. The divergence proceeds until the two groups are different species. Darwin, for example, observed that "offspring of each species will try ... to seize ... diverse places in the economy of Nature .... Each new variety or species, when formed, will generally take the place of ... its less well-fitted parent."[9]

- Niche seeking: Evolution opportunistically fills new ecological niches with life, based on a process of variation that makes a few individuals particularly suited to certain conditions outside the niche occupied by their parents. The species then splits through niche splitting of the newly extended niche, or through heteropatry. Darwin stated a natural result of this process: "The same spot will support more life if occupied by very diverse forms."[10] Furthermore, more species represent more potential hosts of other organisms that depend on them — so a new species is a new niche to be filled by still more new species having parasitic, mutualistic, or commensal relationships with the host species.

- Evolving evolution: Although the genetic code describes the blueprints of an organism, the language of those blueprints is subject to evolutionary change as well. Just as organisms tend to get more fit over time because their blueprints improve, the blueprint description language embedded in the DNA tends to improve as well, making some organisms more evolvable than others. The language also has multiple levels of meaning,[11] just as a passage in English has multiple levels of meaning determined by the words, the intonation, the style, and so on. Species with better evolvability will tend to evolve more efficiently than those with less evolvability, tending to displace them. Thus over long time periods DNA coding strategies with high evolvability will gradually predominate over others, because species with high evolvability will tend to adapt better to changing ecological conditions.

- Improved reproductive strategies: Humans know that finding an ideal mate can be no easy task. Moths have the same problem. Some male moths can detect a single molecule of female scent miles downwind with their feathery antennae, using that clue to fly in her direction; human males are not so sensitive. Good reproductive capabilities become increasingly important when there are few other members of the species available. These can include advanced mate finding methods, ability to store sperm for years after mating (found in some insects, for example), and parthenogenesis (ability to reproduce without mating under appropriate conditions, as in aphids, sea-dwelling foraminifera, komodo dragons, and, many people believe, at least once in humans). Otherwise the process of speciation is limited by the fact that the more species occupy an environment, the fewer members of each species the environment can support. Too few individuals leads to species non-viability when the individuals cannot find mates or other ways to reproduce. This problem is alleviated by improved reproductive strategies. Thus as evolutionary pressure improves reproductive strategies over the eons, species can better prosper with low numbers of members. This means more species can occupy the same space, a process that may continue to gradually improve into the distant future.

- Complex variation: More complicated organisms have more organs, tissues, biochemical processes, homeostatic parameters, genes, and other parts. More parts create more opportunities for variations in offspring because each part can vary. More variations can make evolution work better because there are more potential adaptations to test, and more potential variations means more opportunities for new species to come into existence. This is even more effective than it sounds, because more ways to vary tends to provide more evolutionary paths from an organism's present state to another state that is better adapted to some future, changed environment.

An analogy can provide a general intuition for this: It is harder to get from point A to point B if you must travel in a straight path, because of the obstacles that you could go around if you had many alternative paths. Consequently, when organisms increase in complexity over time they can branch off new varieties, the first step to becoming new species, more readily. In fact there is reason to believe that under many circumstances organisms do tend to increase in complexity over time, though there are also circumstances that create a propensity for simplification (as with bacteria, which appear to follow an evolutionary strategy stressing efficiency over other adaptive strategies).

The natural life cycle of complex human-engineered systems tells us that as we seek to make our engineered artifacts better, they naturally tend to acquire more and more components whose duties are to make them work better by doing more things, things better, and things under more circumstances. For example your cellphone is not only better than the cellphone of yesteryear, it is also more complicated. Organisms can benefit from the same paradigm of increasing complexity over time. The mechanism for this is genetic mutation.

The megatrend toward increasing complexity may apply to the human species as well. With about 25,000 genes, humans have only about 25,000 ways to simplify by deleting a gene and 25,000 ways to complexify by doubling a gene. However the complexification change is more likely to result in a viable organism, because deleting a gene is a bit like removing a random part of a car — likely to cause significant problems. On the other hand, a duplicate copy of a gene can evolve independently of the original, ultimately providing a new and useful function — and making the organism more complex.

Burgeoning biosphere

As future life forms proliferate, the biosphere will become progressively denser — riotous with life. One reason is, to reiterate Darwin's quote above, "The same spot will support more life if occupied by very diverse forms." Another reason is the fuller use of ecological niches that are now only sparsely occupied. These niches pose challenges to life in its present state that should be addressable by advances in evolution or genetic engineering.

Sub-optimal nitrogen content is often a bottleneck in soil fertility. Although the atmosphere is 78% nitrogen, accessing that nitrogen is biochemically challenging and plants cannot do it by themselves. Some plants, however, get around this constraint by hosting bacteria in their roots that do "fix" or extract nitrogen from the atmosphere. The legume family is the most prominent example. This very successful family includes edible beans and many other plants, including various trees. Legumes thus enable an increase in occupancy of low nitrogen habitats. Other nitrogen-fixing plants exist as well. An example is the alder tree. Nitrogen-fixing plants make soil more productive when they decompose by releasing their nitrogen into the soil for other plants to use. There is considerable future potential for nitrogen-fixing plants to speciate into new niches and for non-nitrogen-fixing plants to evolve or be genetically engineering to fix nitrogen as well.

Where drought is the limiting factor, as it often is, future plants may become able to extract water molecules from the air. They could also form communal root conduits extending long distances to distant water sources.

Nutrient-poor ocean water is a bottleneck to ocean surface fertility in many places. But if floating seaweeds would only grow very long descending root threads, they could extract more phosphorous, iron, and other nutrients they need from a larger volume of water when nutrient scarcity is an issue, enabling them to grow in more profusion.

Cold environments could be more densely populated when more organisms there evolve sap and blood that contain better anti-freeze chemicals. Some ice-dwelling organisms already exist. Watermelon snow (colored red or pink by snow algae) is well known, and ice worms that live by eating such algae can burrow through ice, although they essentially liquify at just a few degrees above freezing. The potential is much greater than the meager beginnings the biosphere has now. The frozen wastelands of Antarctica and elsewhere could become covered with vegetation, if not as tall as a rainforest then only because of insufficient nutrients in the icy substrate.

Rock surfaces often have lichens growing on them, but today's rather impoverished lichen ecologies could eventually become much richer, densely covering solid rocks with a vibrant, fast-growing mat of multiple competing lichen species that efficiently erode the rocky surface as they extract the minerals they need directly from the solid rock. Even the barren summit of Mt. Everest could potentially be a jungle of lichen.

Even the sky can be a habitat for life. Birds are nice but the potential for life in the sky goes way beyond our feathered friends. Moisture in clouds, nutrient-containing dust suspended in the atmosphere, and sunlight provide the basis for sustaining tiny photosynthetic microorganisms, whose spores could continue to float high above even when the clouds that produced them dry out. These spores could contain essential minerals that a growing microorganism needs, recycled from the parent cell that dried out earlier. There is thus the potential for a thriving ecosystem in the skies. The current status of life in the sky suggests the plausibility of such a scenario.

Biomass living in clouds today has the "potential to increase by as much as 20% per day"[12] and is estimated to metabolize in the neighborhood of "1 million tons of organic carbon per year."[13] The bacteria *Erwinia caratovora* is just one species that spends part of its life cycle in the clouds.[14]

Some bacterial protein coats are built to encourage ice crystals in the sky to form around them. Some strains of *Pseudomonas syringae*, for example, are coated with proteins containing a "magic" 8 amino acid sequence AGYGSTET which, unlike typical proteins, catalyzes ice crystal formation by aligning water molecules in the required crystalline configuration.[15] Such bacteria are called ice nucleators (INs) and are thought to seed clouds to encourage precipitation.[16] This washes down seed bacteria far from where they started, helping them disperse across the Earth's surface. The significance of this to weather and climate is unknown but perhaps high: clouds typically hold ten thousand microbes per cubic meter.[17] In fact a surprising one third of ice crystals in clouds were found to be nucleated biologically in a test of skies over Wyoming.[18] Sky-dwelling algae could be useful to humans of the distant future, who might transport them to the high clouds of Venus to play a critical role in terraforming an entire planet by converting carbon dioxide to oxygen, because photosynthesis consumes carbon dioxide. From the example provided by our own planet, we now know that a sky teeming with different species of life is not as unprecedented a concept as one might have guessed.

Must all good things come to an end?

A long term increase in the quantity and variety of life, despite the current mass extinction event associated with the human era, is an interesting and perhaps pleasant thought. Ultimately, however, this megatrend will face head winds. These will become more severe over time, suppressing and finally ending life on Earth.

The sun is becoming brighter over time, with changes that become major over a time scale of hundreds of millions of years. On Earth, atmospheric carbon dioxide (CO_2 — the greenhouse gas causing the current global warming problem) helped warm the ancient Earth by high atmospheric levels (and hence a large greenhouse effect) when the sun was dim. Lower levels have helped keep the Earth cool as the Sun has brightened. Very long term, over 100s of millions of years, CO_2 levels are projected to continue decreasing as the increasingly luminous sun heats up the Earth, because CO_2 reacts with common silicate-containing rocks more at higher temperatures, which will drain CO_2 from the atmosphere. This decrease will help keep surface temperatures from rising as quickly as they otherwise would, but will also stress the biosphere because plants need CO_2 for photosynthesis, and very low levels make life harder for plants. This stress will likely start reducing diversity of life[19] at some point, counteracting the trajectory shown in Figure 13.

Eventually, CO_2 levels will become so low that the CO_2 greenhouse warming effect will become negligible. After that, it will be unable to moderate the warming effect of the brightening Sun by lowering further, and the Earth will heat up at an accelerating rate. Too much heat is also a biosphere stressor, and that will also tend to reduce life[20], again counteracting the trajectory of Figure 13.

Two billion years from now, give or take a few hundred million, Earth will become so hot that water will evaporate more, increasing the water content of the atmosphere. High in the sky, water molecules will broken up by high-energy ultraviolet light, and their small, fast-moving constituent hydrogen atoms will permanently flee into space. The Earth's life-giving bodies of water will thus disappear. Surely life on Earth will be ended by then. But if that wasn't enough, in 5.4 billion years the Sun will enter its red giant phase, scorching the Earth as it expands, finally to engulf the Earth in its thin outer atmosphere. The consequent drag will cause the Earth to progressively disintegrate, trailing a long tail of vaporized matter as it spirals, shrinking, into the Sun, as devoid of life as when it formed approximately 4.54 billion years ago.

Instead of hundreds of millions of years, decades

Nature will likely increase life's richness, diversity, and density without our help, given hundreds of millions of years. But with our help the process could happen much, much quicker. At the present high rate of advancement in biotechnology, the speedup could start in a few years and may revolutionize our society in mere decades. This is because genetic engineering could make creating new species a matter of scheduled planning, design, and execution rather than eons of evolutionary trial-and-error. Two helpful contributors to this would be a practical understanding of the language of DNA and advances in hybridizing different species.

A practical understanding of the language of DNA
    The language of genetics is written in the 4-character alphabet A, C, G, and T, which stand for the four nucleobase molecules used to make DNA. How does the sequence of A's, C's, G's, and T's in the DNA of organisms' genomes contribute to the meaning of those genomes? Just like writing down a meaningful sequence of letters requires knowing a human language (English, for example), and creating useful computer software using a sequence of letters and numbers requires knowing a computer language (Java, for example), so creating new biological effects by designing a new DNA sequence

using A, C, G, & T molecules requires understanding the language of genetic coding. We don't know that language very well yet. It is complex and multi-faceted. But we're learning.

Advances in hybridizing different species

Sexual reproduction is popular among higher life forms, such as flowering plants (not to mention birds, bees, and storks). The key selling point from an evolutionary perspective is that it causes mixing of genes, or hybridization. This makes offspring different from their parents. Thus evolutionary competition is not among different versions of a genome (representing different individuals of a species), but among different versions of *parts* of a genome — chromosomes, genes, and intermediate chunks of genetic material consisting of a group of genes in the same region of a chromosome. Without mixing of genetic material during reproduction, these species would change more slowly because offspring would be identical to their parents except for the occasional mutations that would creep in. If species changed more slowly, they would be less able to evolve in response to their environments and thus less able to split into divergent descendant species. Thus, to make such a strategy work well, the genetic coding language would need to permit mutations as frequently as needed to keep the organismal line adapted. It is conceivable such a strategy dominates higher life forms on some mysterious and distant life-bearing planet.

There is another way to mix genetic material. Highly efficient organisms use it, namely the bacteria. The end product of billions of years of evolution (like all other life forms on Earth), bacteria are so efficient in their life processes that under favorable circumstances some can comfortably reproduce as little as 20 minutes after splitting from their parent (which is when they were newborns, in a bacterial sense). In contrast, a human generation of 20 years is 525,960 times longer. Add in four years of college tuition for many human offspring, which bacterial offspring don't need, and it's clear bacteria have honed efficiency into an advanced evolutionary art form. Bacteria thus can rightfully take quiet pride in being top dogs in the hierarchy of life, tolerantly regarding us humans with mild amusement (and perhaps a tinge of derision) as lumbering, jury-rigged contraptions that clearly can't hack it in the race for evolutionary progress into the ranks of advanced, efficient organisms. But let us not dwell on that. Instead, what is the method that bacteria use to mix genetic material, thereby achieving global domination, and how could we use it ourselves?

Bacteria's secret weapon in their race to glorious victory as the world's most advanced life forms is called conjugation. In bacterial conjugation, two bacterial cells either come into contact or form a narrow tube connecting them, and one of them (the donor) transfers genetic material to the other (the recipient). This mechanism is so robust it permits transfer of genes between individuals of different species.[21] A bacterial cell so affected is thus to some degree a hybrid of different species. But species hybridization does not work very well in other kinds of organisms, such as mammals. A mule, for example, is a cross between a horse and a donkey, but mules stubbornly refuse to have offspring themselves except on extremely rare occasions.

However bacteria now have a serious competitor: *Homo sapiens*, i.e., us. Alarmed pundits of the bacterial world might not be fulminating against us in the bacterial blogosphere. Nevertheless we are rapidly improving our ability to do essentially what bacteria can do: transfer genetic material between different species. The most famous early example of this was importing a gene for "green fluorescent protein" (gfp) from a species of jellyfish, *Aequorea victoria*, into other animals. When gfp is present in a tissue, the tissue glows green when exposed to ultraviolet light. Animals that glow green under ultraviolet are so useful to science that Chalfie,

Shimomura, and Tsien received a Nobel Prize in 2008 for being the first to transfer this gene across species.

Glowing animals make fun pets, too. Glowing animals that have been produced include zebrafish (called GloFish), mice (called NeonMice), pigs, and even cats. It could certainly be done with humans too. Perhaps in your lifetime the first celebrity pop star to have a glowing green baby (hopefully only under ultraviolet light) will hog the buzz for a moment in the sun.

Green fluorescent protein is an important early example of cross-species genetic mixing. New examples will occur repeatedly and be quite beneficial. Commodity and specialty crops will undoubtedly be modified to improve their trace nutrient profiles, and with relatively low metabolic cost, making them much more complete foods and benefiting billions of people.

Various animals could be given different human genes, like the FOXP2 gene implicated in human language, to see what the effects would be and thus help elucidate what FOXP2 (for example) actually does in humans. Providing animals with the capacity for language may be a natural next step, revolutionizing human-animal relations. Surely a cat has the cognitive ability to think the concept "mouse smell is here" even if it does not have the ability to say it in words — yet. I predict that dogs have the cognitive, and perhaps soon the linguistic, capacity for interesting conversations about topics that dogs like to think about (like sticks, balls, noises, walks, cats, other dogs... and, of course, scents).

Of course discussions with pets might get repetitive and boring, at least for their human companions. Since conversation might get old with a dog whose experiences and intellectual capacity are necessarily limited compared to our own, why not throw in genes for smarter brains? Of course such genes exist; we are proof by example. One such gene appears to be the SRGAP2 gene, which is partially duplicated in humans, the partial copy being called SRGAP2C.[22] (If you were wondering why the "C," SRGAP2B is another, inactive duplication, while SRGAP2A is proposed as the new name for SRGAP2.) If you add the same duplication to mice (this has already been done), would you get smarter mice? Even more interestingly, intentional duplication might lead to smarter chimpanzees and gorillas, which are already relatively intelligent and human-like in various ways.

Humanzees and humorillas

Species that are fairly closely related, such as humans, chimpanzees, and gorillas, are typically hard to cross. Mules are well known interspecies hybrids but, as noted earlier, generally not fertile themselves. Humanzees (human-chimpanzee hybrids) do not exist, at least not yet. It's been tried. Russian biologist Ilya Ivanov (1870 — 1932) led the effort in the 1920s, though not at the behest of Soviet dictator Josef Stalin as has been colorfully claimed.[23] Unfortunately female chimpanzees were found to be too hard to work with in quantity (only three were tested; identity of human sperm donor not disclosed). So he recruited human female volunteers instead and even claimed to have found one (again, identity was not disclosed). But he ended up never testing her. The three female chimpanzees that were tested did not become pregnant and so the experiment failed. But keep in mind that, because of the small number of tests, such a cross is still a real possibility. Better not try it at home though. Chimpanzees are quite strong, and dangerous when upset.

Whatever the potential may be of getting a human-chimp hybrid by fertilizing an egg cell from one species with a sperm cell from the other, it is more likely that genetic engineering could create a viable hybrid humanzee by other means. To do it, we could try to mix and match chromosomes from one species with those from the other. The beauty of this approach is that the mixing need not consist of exactly half the chromosomes from one with the complementary half from the other, as happens with ordinary egg+sperm fertilization. Instead, any chromosomes from one species could be mixed in with the complementary chromosomes from the other. It seems very likely that there is some combination — or many — that would result in a new kind of being that is both viable and could breed true. One can only imagine what the characteristics of such a new being might be. Would something with two chimpanzee chromosomes out of 46 total be considered human, or at least human enough to have all the rights and privileges appertaining thereto? Or just 44/46 of said rights and privileges? One can only imagine the possibilities. In fact, we probably should imagine them (thus enriching a mostly unrecognized thread in science fiction) because it may anticipate eventual reality.

A burst of new species may result from deliberate genetic engineering, because it is so much faster than natural evolutionary processes and because we can do something nature cannot, which is to combine genes from almost any organisms. That's not exactly evolution as envisioned by Darwin, in which the most important mode of genetic change, random variation with selection, is very slow and takes place in indistinguishably tiny steps, a process that can lure species into the local maximums of evolutionary cul de sacs.

Ultimately the more kinds of organisms there are, the more genetic opportunities exist for new species to branch off, thereby speeding up evolution in general and speciation in particular.

Chapter Thirty-One

If the Universe As We Know It
Ends, When Will It Happen?

Space is mostly vacuum, but vacuum is not really "empty." In fact vacuum might really be "false vacuum," which could transition to a lower energy state, ending the universe as we know it. If that is destined to happen, it probably won't be soon. But it definitely might be.

The universe might not end unexpectedly, which is good. But then again, it might — that's bad. This could happen from what physicists call a *vacuum metastability event*.

It turns out that all of space, even where there is complete vacuum, might be stuck in a high energy state. Seemingly stable since time immemorial, it might actually be stable. Such a high energy state for a vacuum is termed a "false vacuum." We don't know if vacuum in our universe is false (i.e., high in energy) or not, although we do know it is not truly empty — it is actually full of ephemeral virtual particles that spring into existence and then quickly out again.

If vacuum in our universe is false, the state of some point somewhere in the universe could one day suddenly transition to a lower energy state. This could happen, for example, as a result of random quantum fluctuation somewhere in the universe. If such a transition occurred, crazy things would happen fast. The physical laws governing the universe would change at that spot, radically changing or simply destroying whatever happens to be there. This might sound like no big deal, as long as you're not in the immediate vicinity. However, it actually is a big deal because the lower-energy state of this spot would spread outward at almost the speed of light. The changed physical laws, annihilation of matter, or whatever it is that is associated with this lower energy state would consequently spread outward at the same rate. If it started on Earth, the Earth would be destroyed more or less instantly. If it started anywhere else in the universe, the spreading destruction would arrive here, traveling outward at the speed of light in all directions from the starting point. Once it arrived, the Earth as we know it would be gone in the blink of an eyelid.

We know little about what this lower energy state of the universe would be like, but physicists have theorized a bit. A study by Coleman and de Luccia concluded that, heretofore, "one could always draw stoic comfort from the possibility that perhaps in the course of time the new vacuum would sustain, if not life as we know it, at least some structures capable of knowing joy. This possibility has now been eliminated."[1] Bummer!

If this seems hard to wrap your mind around, you are not the only one. Even the physicists who investigate the mathematical models from which they draw such conclusions can't always tell if their equations apply to reality or not. For example, faster than light travel seems possible if speed is described using equations that permit going "around" the barrier presented by the exact speed of light.[2] But that doesn't mean the equations hold — they might, or they might not.

As pointed out by philosophers of science like Karl Popper and others, scientific theories cannot be proven absolutely.[3] Instead, evidence builds up in favor of, or against, a theory and a rational person goes with the evidence. Confirmation of whether the universe is unstable or not awaits more data. In particular, we need to know the precise masses of two particles. One is the Higgs boson, sometimes called the "God particle" to the chagrin of many experts, central though it is to our understanding of the universe. The other is the "top quark," a type of, uh, quark, a name borrowed from James Joyce's novel *Finnegans Wake*, in which a seagull flying overhead cries "three quarks for Muster Mark!"

Since we don't know the required masses accurately enough, we don't know if the state of the universe is a false vacuum and, if it is, whether it will ever transition out of it and thus destroy everything we are familiar with.[4]. So maybe we have nothing to worry about, but maybe we do. Just how much should we worry? If such an event will at some point end us, clearly a key question is when. If it happens in billions of years, there seems little point in breaking a sweat over it now. On the other hand if it might happen soon, perhaps you should start eating dessert first.

Interestingly, though the detailed physics of a vacuum metastability event are mostly impenetrable to anyone but a specialist, the average reader is capable of understanding when it (or some other unpredictable end to the universe) will occur, if it does. Its very unpredictability is the key to when it might happen.

Guesstimating when

Glance at your watch. There is a 50% chance the seconds reading will be between 0 and 30 (i.e. in the first half of a minute) and a 50% chance it will be above 30 (in the second half). Now let's extend this idea in an understandable way. There is a 1/60 chance that the moment you start reading this is somewhere within the first 1/60 of the current minute and that 59/60 or more of the minute remains. We know a minute is 60 seconds long, so the first sixtieth of it spans the first second and the 59/60 of it that is left is 59 seconds long. Similarly there is a ¾ chance that this moment is within the first ¾ of a minute (the first 45 seconds) with ¼ minute (15 seconds) or more left, and so on. If you're with me so far, hang in there because things are about to get more interesting — but still understandable.

Suppose you are dealing not with a minute but with a time period of unknown length — say the lifespan of our little area of the universe, defined as the time before an expanding metastability event ends the Earth. Just as in the case of a minute, there is a 1/60 chance that the present moment is within the first 1/60 of this time period, which is the lifespan of our area instead of a lowly minute. Continuing the analogy, there is a 1/60 chance that 59/60 or more of the lifespan of our area remains. Since the universe is 13.82 billion years old, there is a 1/60 chance that 59/60 or more of our area's lifespan remains, in which case 59×13.82 billion = 815.38 billion years or more are still more or less safely ahead of us. This may sound reassuring, but note that this has only a 1/60 chance. There is thus a 59/60 chance — over 98% — that, assuming the universe ever starts disintegrating at an unpredictable moment, the disintegration reaches us in *less* than 815.38 billion years. Maybe a lot less. Let's explore this admittedly unappealing possibility next, taking at least "stoic comfort" in the knowledge that even if a vacuum metastability event destroys the universe, it will at least have had the privilege of containing "structures capable of knowing joy" in its history.

To start, assume for purposes of argument that at some point there will be a metastability event. That's a big assumption to make, and it might or might not be true, but the analysis only works if we make it. Then, a randomly chosen point within the lifespan of our area has a 50% chance of being in the first half of that lifespan and a 50% chance of being in the second half (if the probability of a metastability event is constant over time; if not, see the reference[5]). Indeed today, the day you read these words, is such a random point in time. Recall that the universe is known to be currently 13.82 billion years old, give or take a small matter of a hundred million years or so. Those 13.82 billion years thus have a 50% chance of being in the first half of the life, in which case the solar system will last at least another 13.82 billion years, for a total of 27.64 billion years or more. But don't breathe a sigh of relief just yet: The other 50% chance is associated with our being in the second half of our lifespan instead of the first. In that case, there is less than 13.82 billion years left, since if we're in the second half of our time, the remaining time must be less than the already-elapsed time of 13.82 billion years. So can you breathe easy, or not? To decide this question, we need to extend the analysis a bit further.

We've already done a 50% / 50% split of the life, so let's try another one: 75% / 25% (see Figure 14). From the perspective of this split, there is a 75% chance that today, a random day within an unpredictably ending lifespan, will be in the first 75% of that span.

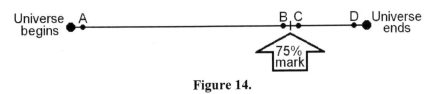

Figure 14.

Suppose today is point A in Figure 14. Then A represents the 13.82 billion year point, and things are just getting started. Our part of the universe has a long, long future before reaching its end point. With a ruler and calculator one could calculate just how long that future is from the time line in the figure (it turns out to be about another 290 billion years). But points B, C, and D (as well as all the other unmarked points on the time line) are equally as likely to represent today as point A. So suppose instead that point B is today. In this case the lifespan is almost 75% over, and by the same reasoning as for point A, we're 13.82 billion years into a 18.43 billion year lifespan, giving us 4.61 billion years more. While this is a lot less than if today is at A, it is still way too far off to panic about. We thus conclude that there is a 75% chance of nothing to worry too much about, but what about the remaining 25%? This corresponds to now being near point C, near point D, or somewhere between C and D in the figure. The question then becomes, how long are the shortest 25% of the set of all possible lifespans? This is important because, while there is a 75% chance the lifespan is longer than that (whew!), a 25%

chance for it to end sooner is still pretty high, considering it's existence itself (or at least its ability to support joy) that we're talking about.

Focusing on the 25% chance that today, a random point in time, is in the last 25% of the lifespan, the first 75% must therefore be less than the 13.82 billion year current age of the universe. See Figure 15.

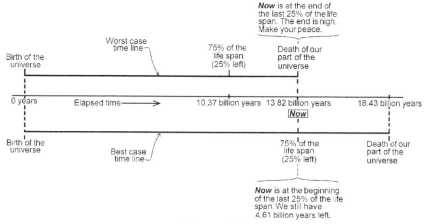

Figure 15.

This figure shows bolded lines illustrating the shortest (top) and longest (bottom) lifespans, if today is in the last 25% of the lifespan. From the figure, we can conclude that there is a 25% chance of the universe as we know it ending within the next 4.61 billion years, assuming it does indeed end unpredictably (which again, it might or might not do). The worst case is that it ends within the next few minutes, but that sounds rather unlikely considering the billions of years of other possible times. But how unlikely is an uncomfortably soon end? For that we need to go beyond the 75/25 perspective.

We've discussed a 50/50 split and a 75/25 split. How about a 99/1 split? That split highlights the fact that there is a 1% chance we are in the last 1% of the life of our spot in the universe. Then the same type of reasoning behind the 75/25 analysis applies, except this time the best case is a lifespan that lasts another 140 million years, or 1% of a total of 13.96 billion years of which the current 13.82 billion years elapsed is 99%. Now 140 million years is still a long time although it's getting short enough to give some pause — dinosaurs roamed the planet 140 million years ago, for example. Nevertheless, 140 million years is not exactly tomorrow.

What about a split that creates a time frame that one might actually worry about? Say, one that could induce you to think seriously about eating dessert first? Here is the straight dope: If the universe as we know it is destined to end at a random and unpredictable time, there is one chance in 100 million it will be within the next 138.2 years and one chance in a billion it ends within the next 13.8 years.

You heard it here first.

Recommendations

Feel free to continue enjoying dessert in its traditional role of a post-main course treat. While 138.2 years (or 13.8) is soon enough to be of concern, the probability level of 1 in 100 million (or 1 in a billion) is probably not. Your chances of being hit by lightning, for example, are many thousands of times greater, roughly 1 in 10,000, making it a much greater concern.[6] Take precautions during storms![7] Seek shelter, but not under a tree as lightning may target the tree. Long, thin metal objects such as umbrellas and golf clubs also may tend to attract lightning, particularly when raised above the head. If you feel your hair standing on end, you have become electrically charged and lightning may strike. Crouch down close to the ground immediately, because lightning seeks targets that are higher than the ground.

Interestingly, the same type of analysis we have described can be applied to human events with more ordinary time scales. For example, suppose July 4, 2020 rolls around with the universe still more or less intact. The United States, born in 1776, will be 244 years old.

Assuming July 4, 2020 is a random day in the lifespan of the US, there is a 5% chance that 244 years is at least 95% of the country's lifespan and thus that the country will end by just under 13 years later, in 2033. By a similar analysis, there is a 1% chance it will end by a week or two shy of 2½ years later, in 2023. Think that's crazy? As I write this, a large fireball is in the news. It appeared last night in the skies over portions of the Midwest US, accompanied by a loud sonic boom. Probably caused by a meteoroid, it broke up into smaller fireballs before disappearing. Although it caused no damage, had it been big enough, it would have devastated the land, ending the aforementioned lifespan, or even devastating the entire Earth (as happened to the dinosaurs).

Other scenarios in which the US ends are caused by humans, not meteors from the sky. Considering the turmoil the world has seen in various locations in the recent and less recent past, and the analysis above, it does seem risky to assume long-term stability in the US. For example, the presidential democracy system used in the US has been unstable when tried in other countries. This suggests that the US as we know it might be (probably is?) less stable than we might think. Thus, a clearly prudent recommendation is to take steps to ensure the stability of the system of government that Americans proudly consider an essential ingredient of what makes the US the US.

But what about Canada? If you're Canadian, your country was born in 1867. There is a 5% chance the end would be within 8 years and a couple of weeks after 2020 and a 1% chance it would be within just over a year and a half of that time (assuming 2020 is reached successfully). The apparent greater stability of the US (deduced from the historical trend of already having existed for longer) gives it a greater than even chance of outliving Canada. What is that chance? In the case where the lifespans of the two countries are independent, a mathematical technique called convolution can actually compute the chance that the US outlasts Canada: about 58%. On the other hand,

Canada's corresponding probability of outliving the US is 42%, not that much less.

Now you know.

Hopefully, 2020 is not a random year in these national lifespans. For example, national lifespans might be tied to the lifespan of western civilization, which started much longer ago and would thus lead to longer estimates of remaining life. That is certainly possible — but there's no guarantee. As another complicating factor, there is actually a variety of apocalyptic events that can occur, and they should all be appropriately analyzed to determine how concerned we should be.[9] Because we are in the "Recommendations" section, I suggest careful scientific, mathematical, and philosophical study of this important issue, funded by the US and Canadian governments — and other countries all of which have analogous predicaments.

Maybe eating dessert first is not a bad idea.

Chapter Thirty-Two

Questions

On time and the universe…
meaninglessness and meaning.

Will the universe as we know it, with humankind in it, disappear? See chapter 30. Billions of years from now will our comfortable Big Bang universe reverse course, ending up as a giant black hole in a "Big Crunch," perhaps followed by another big bang called the "Big Bounce"? Current data does not support this fate. Or will our expanding universe accelerate outward, with galaxies, planets, sentient beings, and ultimately our very atoms torn asunder, their mangled shreds blasting outward for eternity in a "Big Rip"? Although the expansion *does* appear to be currently accelerating, the prediction favored by most cosmologists is not the Big Rip but rather continued expansion and cooling, the universe ultimately sliding into a "Big Freeze."[1] Apparently, we do not know specifically what will happen, but we do know it will be Big.

Does the universe really exist, or is it instead being simulated inside a computer (or something), which is part of another universe that we cannot see, just as a character in *The Sims* cannot see our universe, which in turn contains it? The concept of a complex world inside of a computer became popular with movies like *Tron* (1982) and more recently *The Matrix* (1999) and its sequels and derivative comics, animations and video games. That our own "real" universe could be more likely a simulation than not has been seriously argued in the philosophy field, and experiments capable of providing evidence for or against are described by physicists.[2]

What about us humans? Will we someday meet our match among the stars, or will the famous Fermi paradox — the apparent lack of the other intelligent races predicted by the equally famous Drake equation — be our future instead? If you believe in UFOs, we've already met at least one civilization more advanced than our own. If not, then Fermi's paradox applies: Where are they?

Will we destroy ourselves in a matter of years, decades, or centuries, or will we last indefinitely? Many people see our increasing power over nature, coupled with self-destructive tendencies, leading us to destroy ourselves. On the other hand, if more typical evolutionary processes dictate our fate, it is known that most species last no longer than a few million years.[3] Over that period most species either go extinct or change into different species. Thus our distant descendants may not exist; and if they do exist they would likely not be human as we recognize it today — resembling us, their forebears, as little as we resemble our own forebears the apes, the first mammals, the fish, or going back even further, the roundish flatworm.

Perhaps we will transcend both technology-enabled self-destruction and normal evolution. What then? Will we change ourselves by genetic engineering and until our descendants are... something, but not human as we know it today? This is plausible because biotechnology is advancing fast.

Those questions are important. Or are they? Their answers have little if any impact on our lives. But, at least the questions have *existential* importance. Or do they? After all, that huge ball of rock and iron thousands of miles thick we call the Earth is indifferent to our existence. Our activities, literally, barely scratch the surface.

However, questions can still be important. Our brains are highly complex systems, capable of suffering and enjoyment, emotion, a sense of purpose and meaningfulness — emergent properties arising from the complexity of a system that, like all emergent properties, help make systems more than the sum of their parts. Such mental states intrinsically lay persuasive claim to their own importance, implying that questions deriving from them have great importance. A pre-eminent such question then must be, how can we maximize enjoyment in the world or the universe, minimize suffering, and measure progress toward that goal?

Other questions clamor as well. Will our occupancy of a tiny slice of eternity change the distant future as the flapping of a fragile butterfly's wings changes the course of hurricanes years later (chapter 16)? Or will our existences as paupers, kings, and even as a species fade away like a sunspot, something that happened, but then vanished, leaving no detectable trace in the vastness?

References and Further Reading

Introduction: Why Read This Book?

1. "Prediction is very difficult, especially about the future," B. Popik, *The Big Apple*, Oct. 30, 2010, www.barrypopik.com/index.php/new_york_city/entry/never_make_for ecasts_especially_about_the_future.
2. "The future is not what it used to be." quote attributed to L. Riding and R. Graves, 1937, by G. O'Toole, *Quote Investigator*, quoteinvestigator.com/2012/12/06/future-not-used.

Chapter One: What It Means That an Hour's Work Yields a Week's Food

1. "But by 2000, just 1.9% had agricultural jobs, more than a 20-fold improvement in productivity per person." C. Dimitri, A. Effland, and N. Conklin, The 20th century transformation of US agriculture and farm policy, Electronic Information Bulletin Number EIB-3, US Dept. of Agriculture Economic Research Service, June 2005, www.ers.usda.gov/publications/eib-economic-information-bulletin/eib3.aspx.
2. *La Dolce Vita* is a movie directed by Federico Fellini, 1960.
3. *Economic possibilities for our grandchildren*: J. M. Keynes, reprinted in his *Essays in Persuasion*, W. W. Norton & Co., 1963, pp. 358–373, www.econ.yale.edu/smith/econ116a/keynes1.pdf.
4. "In 2013 the Netherlands had an average work week of just 29 hours ...": E. Lewis, The Netherlands has the shortest work week in the world! *IamExpat*, July 27, 2013, www.webcitation.org/6St5WPw7N.

5. "Taiwan reduced its work week to 42 hours in 2000," Taiwan legislature approves 42-hour workweek, *Asian Economic News*, June 16, 2000, www.webcitation.org/6ZYt1TLxH.

6. "Similarly, the cost of artificial light has also decreased dramatically over time." W. D. Nordhaus, Do real-output and real-wage measures capture reality? The history of lighting suggests not. Chapter in T. F. Bresnahan and R. J. Gordon, eds., *The Economics of New Goods*, U. Chicago Press, 1996, www.nber.org/chapters/c6064.pdf.

7. "Even billionaires think people work too much." A. Newcomb, Billionaire Richard Branson wants everyone to have unlimited vacation, ABC News, September 23, 2014, www.webcitation.org/6SshDkcHb.

8. "Milton Friedman, who advised both Ronald Reagan and Margaret Thatcher," C. Moore, *Not for Turning*, vol. one of *Margaret Thatcher: the Authorized Biography*, Penguin, 2013, pp. 576-7.

9. "Richard Nixon attempted (but failed) to implement a plan based on Friedman's." N. Gordon, The conservative case for a guaranteed basic income, *The Atlantic*, Aug. 6, 2014, www.theatlantic.com/politics/archive/2014/08/why-arent-reformicons-pushing-a-guaranteed-basic-income/375600.

10. "… only reducing work time by a little, mostly among new mothers and teenagers." E. L. Forget, The town with no poverty: Using health administration data to revisit outcomes of a Canadian guaranteed annual income field experiment, U. Manitoba, 2011, public.econ.duke.edu/~erw/197/forget-cea%20(2).pdf.

Chapter Two: *Smart Pills'n Such — Cognitive Enhancement the Easy Way*

1. "… alpha-CaM kinase II …": L. Gravitz, Selectively deleting memories, *Technology Review*, Oct. 22, 2008, www.technologyreview.com/news/411076/selectively-deleting-memories.

2. "The common anti-diabetic drug metformin, for example, can make mice learn water mazes better; they even grow more neurons in the process." J. Wang, D. Gallagher, L. DeVito, G. Cancino, D. Tsui, L. He, G.

Keller, P. Frankland, D. Kaplan, F. Miller, Metformin activates an atypical PKC-CBP pathway to promote neurogenesis and enhance spatial memory formation, *Cell Stem Cell* [sic], 2012, vol. 11, issue 1, pp. 23–35, www.cell.com/cell-stem-cell/retrieve/pii/S1934590912001749.

3. "Table. Some cognitive enhancement substances and activities." For entries caffeine through methylphenidate, review and additional references provided in A. Sandberg and N. Bostrom, Converging cognitive enhancements, *Annals of the New York Academy of Sciences*, 2006, vol. 1093, pp. 201–227, www.nickbostrom.com/papers/converging.pdf.

For hydromel, an ancient recipe is: Pour some water (hydro) and honey (mel) into a container, mix well, enjoy.

For modafinil, armodafinil is a newer drug available as nuvigil but not available as a generic.

For piracetam and Hydergine, simply start with a web search.

For donepezil, see M. S. Mumenthaler, et al., Psychoactive drugs and pilot performance: A comparison of nicotine, donepezil, and alcohol effects, *Neuropsychopharmacology*, July 2003, vol. 28, no. 7, pp. 1366–1373, www.nature.com/npp/journal/v28/n7/full/1300202a.html.

For cortactin, see: UBC researchers' discovery could rejuvenate the brain, The University of British Columbia, Dec. 2008, news.ubc.ca/2008/12/18/archive-media-releases-2008-mr-08-166.

For magnesium threonate, see I. Slutsky et al., Enhancement of learning and memory by elevating brain magnesium, *Neuron*, Jan. 28, 2010, and www.physorg.com/news183818175.html. Ordinary magnesium supplements ("milk of magnesia," dolomite, etc.) may help as well.

For insulin-like growth factor 2, see D. Y. Chen, et al., A critical role for IGF-II in memory consolidation and enhancement, *Nature*, Jan. 2011, vol. 469, pp. 491–497.

For metformin, see Wang, et al. (2012), detailed above.

For exercise, see (1) H. van Praag, et al., Running enhances neurogenesis, learning, and long-term potentiation in mice,

Proceedings of the National Academy of Sciences (*PNAS*), 1999, vol. 96, no. 23, pp. 13427–13431; (2) P. S. Eriksson, et al., Neurogenesis in the adult human hippocampus. *Nature Medicine*, 1998, vol. 4, pp. 1313–1317; and (3) www.bakadesuyo.com/2014/01/get-smarter.

For afternoon napping, see e.g., (1) A midday nap markedly boosts the brain's learning capacity, www.physorg.com/news185948338.html. (2) E. J. Wamsley, et al., Dreaming of a learning task is associated with enhanced sleep-dependent memory consolidation, *Current Biology*, Apr. 22, 2010, vol. 20, issue 9, pp. 850–855, www.ncbi.nlm.nih.gov/pmc/articles/PMC2869395.

For NgR1 antagonist, see: (1) regarding myelination, J. K. Relton, J. Li, and B. Ji, Use of Nogo Receptor-1 (NGR1) antagonists for promoting oligodendrocyte survival, US patent application US2011123535, May 26, 2011, www.google.com/patents/EP2023735A2; (2) regarding synapses, F. Akbik, S. M. Bhagat, P. R. Patel, W. B. J. Cafferty, and S. M. Strittmatter, Anatomical plasticity of adult brain is titrated by Nogo Receptor 1, *Neuron*, vol. 77, pp. 859–866, 2013, dx.doi.org/10.1016/j.neuron.2012.12.027.

For fish, see: J. Hamblin, This is your brain on fish, *The Atlantic*, August 7, 2014, www.theatlantic.com/health/archive/2014/08/this-is-your-brain-on-fish/375638.

For klotho, see: Anti-aging gene also enhances cognition, *Kurzweil Accelerating Intelligence News,* May 12, 2014, www.kurzweilai.net/anti-aging-gene-also-enhances-cognition.

For GLYX-13, see: (1) J. L, Reilly, Effects of GLYX-13 on learning and memory in healthy individuals and those with psychiatric illness, National Institutes of Health, updated Oct., 2015, clinicaltrials.gov/ct2/show/NCT01844726; (2) Ketamine cousin rapidly lifts depression without side effects, *Science Update*, 6th paragraph, May 13, 2013, www.nimh.nih.gov/news/science-news/2013/ketamine-cousin-rapidly-lifts-depression-without-side-effects.shtml.

4. "This is not electroshock therapy, which applies hundreds of times more electric current (usually 800 milliamps, as much as many lightbulbs, compared to the 1–2 milliamps typical of TES);" 800 mA: MECTA

Corporation – electroconvulsive therapy (ECT) products, www.mectacorp.com/products.html. 1–2 mA: R. C. Kadosh, Using transcranial electrical stimulation to enhance cognitive functions in the typical and atypical brain, *Translational Neuroscience*, 2013, vol. 4, no. 1, pp. 1–14.

5. "Things have come a long way ...": Field reviewed in (1) S. Zaghi, M. Acar, B. Hultgren, P. S. Boggio, and F. Fregni, Noninvasive brain stimulation with low-intensity electrical currents: Putative mechanisms of action for direct and alternating current stimulation, *Neuroscientist*, June 2010, vol. 16, no. 3, pp. 285–307, www.fisherwallace.com/uploads/Harvard_Medical_School_Research.pdf; (2) R. C. Kadosh, 2013; and (3) blog with updates at www.diytdcs.com.

6. "... Sylvanus Thompson's classic electrical engineering textbook described connecting a battery to the forehead to cause a 'wild rush of colour' ...": *Elementary Lessons on Electricity and Magnetism, Fourth Ed.*, ch. 3, lesson XIX, Macmillan and Co., 1883. An edition was reprinted in 2011.

7. "For example tDCS was shown in 2011 to induce insight in tricky problem-solving situations." R. P. Chi and A. W. Snyder, Facilitate insight by non-invasive brain stimulation, *PLoS ONE*, 2011, vol. 6, no. 2, www.plosone.org.

8. "You can even build your own tES machine on the cheap from do-it-yourself plans." Build a Human Enhancement Device (Basic tDCS Supply), www.instructables.com, www.webcitation.org/6ZZ3464o1.

9. "Indeed, in September 2011 DBS finally made mice smarter." S. Stone, C. Teixeira, L. DeVito, K. Zaslavsky, S. Josselyn, A. Lozano, and P. Frankland, Stimulation of entorhinal cortex promotes adult neurogenesis and facilitates spatial memory, *The Journal of Neuroscience*, Sept. 21, 2011, vol. 31, no. 38, pp. 13469–13484, www.jneurosci.org/content/31/38/13469.

10. "Luckily, progress in brain scan technology is improving exponentially." R. Kurzweil, *The Singularity is Near*, Penguin Books, 2005, pp. 159–160.

11. "Another method uses ultrasound ...": W. J. Tyler, Y. Tufail, M. Finsterwald, M. L. Tauchmann, E. J. Olson, and C. Majestic, Remote excitation of neuronal circuits using low-intensity, low-frequency ultrasound, *PLoS One*, Oct. 29, 2008, vol. 3, no. 10, dx.doi.org/10.1371/journal.pone.0003511.

12. "The nanomechanical effects of the ultrasonic sound waves ...": W. D. Jones, Sound waves for brain waves, *IEEE Spectrum Online*, Jan. 2009, spectrum.ieee.org/jan09/7097.

13. "Its connectivity to other parts of the brain explains 10% of the variation in intelligence in humans." M. W. Cole, et al., Global connectivity of prefrontal cortex predicts cognitive control and intelligence, *The Journal of Neuroscience*, June 27, 2012, vol. 32, no. 26, pp. 8988–8999, www.jneurosci.org/content/32/26/8988.

14. "By displaying to the user their brainwave frequencies ... they can actually learn to change them ...": C. Chase and G. Yonas, Cognitive enhancement using feedback, US Patent Application 20150297108, Oct., 2015, www.freepatentsonline.com/y2015/0297108.html. For some background see B. van der Kolk, *The Body Keeps the Score: Brain, Mind, and Body in the Healing of Trauma*, Penguin Books, 2014.

15. "Some evidence suggests that this can 'enhance attention, increase overall intelligence, relieve short-term stress, and improve behavior.'" T. L. Huang and C. Charyton, A comprehensive review of the psychological effects of brainwave entrainment, *Alternative Therapies in Health and Medicine*, 2008, vol. 14, no. 5, pp. 38–50.

16. "... roughly $15,000 per IQ point ...": S. D. Grosse, et al., Economic gains resulting from the reduction in children's exposure to lead in the United States, *Environmental Health Perspectives*, June 2002, vol. 110, no. 6, pp. 563–569 (see Table 2), www.ncbi.nlm.nih.gov/pmc/articles/PMC1240871.

17. "Iodine deficiency ... can significantly lower the IQ of whole populations": The scope of the problem, Micronutrient Initiative, micronutrient.org/what-we-do/by-micronutrient/iodine/. See also N. D. Kristof, Raising the world's IQ, *The New York Times*, Dec. 4, 2008,

New York edition p. A43,
www.nytimes.com/2008/12/04/opinion/04kristof.html.
18. "However, taking choline …": Sandberg and Bostrom (2006), pp. 207, 223, detailed above.
19. "… take cod liver oil …": I. B. Helland, et al., Effect of supplementing pregnant and lactating mothers with n-3 very-long-chain fatty acids on children's IQ and body mass index at 7 years of age, *Pediatrics*, Aug. 2008, vol. 122, no. 2, pp. e472–e479, pediatrics.aappublications.org/cgi/content/full/122/2/e472.
20. "Sometimes called the "critical mass" effect, anyone who has been in such a group can vouch for its effectiveness …": E.g. A. W. Woolley, C. F. Chabris, A. Pentland, N. M. Hashmi, and T. W. Malone, Evidence for a collective intelligence factor in the performance of human groups, *Science*, 2010, vol. 330, no. 6004, pp. 686–688, www.sciencemag.org/content/330/6004/686.

Chapter Three: Keys and Screens
Today, Mind Reading Tomorrow

1. "A brain wave reading device that enables paralyzed individuals to switch lights and appliances on and off without moving at all was reported ready for marketing as early as 1997." Remote-control system uses brain waves, Associated Press, Dec. 25, 1997, goo.gl/uxuAEp.
2. "This device, the MCTOS Brain Switch, was still listed for sale as of late 2015." E.g. Synapse Adaptive, www.synapseadaptive.com/switches/brain%20switch.htm.
3. "In 2008 a team of researchers at Keio University," June 2, 2008, www.webcitation.org/6ZYwA0UZp.
4. "By 2013, non-invasive brain wave detection enabled people to control quadrotor air vehicles by imagining hand motions." University of Minnesota researchers control flying robot with only the mind, *UMNews*, 2013, www1.umn.edu/news/news-releases/2013/UR_CONTENT_445216.html.
5. "Beer finally entered the picture in 2015 …": E. Zolfagharifard, The heartwarming moment quadriplegic picks up his own beer for the first

time in 13 years - using a robot arm controlled by his thoughts, *Daily Mail*, May 21, 2015, u.to/_QSACw.

6. "For the number keys, this principle was demonstrated in 2009," E. Eger, V. Michel, B. Thirion, A. Amadon, S. Dehaene, and A. Kleinschmidt, Deciphering cortical number coding from human brain activity patterns, *Current Biology*, 2009, vol. 19, issue 19, pp. 1608–1615, www.sciencedirect.com/science/article/pii/S0960982209016236.

7. "highly significant accuracies": T. M. Mitchell, S. V. Shinkareva, A. Carlson, K. Chang, V. L. Malave, R. A. Mason, M. A. Just, Predicting human brain activity associated with the meanings of nouns, *Science*, May 30, 2008, vol. 320.

8. "A similar capability was also demonstrated with viewed pictures rather than imagined words," Mind reading machine knows what the eye can see, *New Scientist*, Mar. 2008, www.newscientist.com/article/dn13415-mindreading-machine-knows-what-the-eye-can-see.html.

9. "There is no natural barrier from what we can see": Reported by Agence France-Presse (AFP), July 7, 2008, www.webcitation.org/6ZYwY8lrV.

10. "For example, ancient catapults used... human hair!" J. G. Landels, *Engineering in the Ancient World*, University of California Press, 1978.

11. "Soldiers would need to think in 'clear, formulaic ways ... similar to how they are already trained to talk.'" L. Zyga, US Army invests in 'thought helmet' technology for voiceless communication, PhysOrg.com, Sept. 22, 2008, www.physorg.com/news141314439.html.

12. "... an off-the-shelf brain wave reader was demonstrated as part of a cellphone device ...": K. Grifantini, Mobile phone mind control, *Technology Review Editors' Blog*, Mar. 31, 2010, www.technologyreview.com/view/418258/mobile-phone-mind-control.

13. "... a technique for reading the mind and outputting a fuzzy movie ...": *New Scientist*, Oct. 28, 2009, issue 2732, www.newscientist.com/article/mg20427323.500-brain-scanners-can-tell-what-youre-thinking-about.html.

14. "... a 2013 project read interpretations of Beethoven from peoples' brains and played the results ...": Peninsula Arts Contemporary Music Festival, Feb. 22–24, 2013, www.webcitation.org/6ZYwwDyoa.

15. "... Wright's, Goddard's, and hybrid laws (applying to many technologies)." B. Nagy, J. D. Farmer, Q. M. Bui, and J. E. Trancik, Statistical basis for predicting technological progress, *PLOS ONE*, 2013, vol. 8, issue 2, journals.plos.org/plosone/article?id=10.1371/journal.pone.0052669.

16. "... spatial resolution of non-destructive brain scan technology doubles about every six years, while time resolution doubles about every year and a half." derived directly from figures on pp. 159–160 of R. Kurzweil, *The Singularity is Near*, Penguin Books, 2005.

Chapter Four: **Wiki-Wiki-Wikipedia**

1. "... web inventor Tim Berners-Lee ...": (1) Information Management: A Proposal, CERN, 1989, www.w3.org/History/1989/proposal.html; (2) "The website of the world's first-ever web server," info.cern.ch.

2. "The World-Wide Web (W3) was developed to be a pool of human knowledge." T. Berners-Lee, R. Cailliau, A. Luotonen, H. F. Nielsen, and A. Secret, The World-Wide Web, *Communications of the ACM*, Aug. 1994, vol. 37, no. 8, pp. 907–912.

3. "overflowed this original goal to become a vast sea": N. Montfort and N. Wardrip-Fruin, "54. [Introduction] The World-Wide Web," in Wardrip-Fruin and Montfort, eds., *The New Media Reader*, MIT Press, 2003, p. 791, ISBN 0-262-23227-8.

4. "Engelbart, in the 1960s, built the first hypertext system ...": Called NLS, it grew out of: D. C. Engelbart, Augmenting human intellect: A conceptual framework, SRI Summary Report AFOSR-322, SRI Project No. 3578, Oct. 1962, www.webcitation.org/6ZYx5E9lv.

5. "Engelbart also invented the mouse!" See images at images.iop.org/objects/ccr/cern/40/10/24/cernbooks2_12-00.jpg and en.wikipedia.org/wiki/File:Firstmouseunderside.jpg.

6. "long, long hallway, ... almost featureless" and subsequent quote: W. Van Winkle, Back Door: Q&A with Douglas Engelbart, *Computer Power User*, April 2002, vol. 2, issue 4, p. 108. For similar quotes see other sources such as: Engelbart on the epiphany: "Bingo: It just occurred to me," section in V. Landau and E. Clegg, *The Engelbart Hypothesis:*

Dialogs with Douglas Engelbart, 2nd ed., 2009, NextPress, 2nd ed., ISBN 978-0615308906.

7. "Vannevar Bush's famous 1945 *Atlantic Monthly* article, *As we may think*, ...": July 1945, www.theatlantic.com/magazine/archive/1969/12/as-we-may-think/3881.

8. "Ted Nelson coined the term hypertext ... at Vassar College ...": (1) L. Wedeles, Prof. Nelson talk analyzes P.R.I.D.E, *Vassar Miscellany News*, Feb. 3, 1965, faculty.vassar.edu/mijoyce/MiscNews_Feb65.html. (2) M. Joyce, Did Ted Nelson first use the word "hypertext" at Vassar College? What is Vassar's claim? See faculty.vassar.edu/mijoyce/Ted_sed.html.

9. "His Xanadu project ...": Home pages at xanadu.com.au and xanadu.net.

10. "The World Wide Web is what we were trying to prevent." S. Ditlea, Xanadu's creator at 60: Still visionary, still cantankerous, *New York Times*, June 21, 1997, www.nytimes.com/library/cyber/week/062297nelson.html.

11. *Literary Machines*, T. Nelson, Mindful Press, numerous editions 1980–1993.

12. *It is Earlier Than We Think*: Chapter IX of V. Bush, *Science is not Enough*, Morrow, 1967.

13. "The original general purpose computer was the Analytical Engine of Charles Babbage ...": E.g. C. Babbage, Of the Analytical Engine, ch. 8 in *Passages from the Life of a Philosopher*, 1864, www.fourmilab.ch/babbage/lpae.html.

14. "Programs were nevertheless written for it in 1842–3 by Ada Lovelace ...": Ada Augusta, Countess of Lovelace, English translation with extensive notes of: L. F. Menabrea, Sketch of the Analytical Engine invented by Charles Babbage, Bibliothèque Universelle de Genève, Oct. 1842, no. 82; www.fourmilab.ch/babbage/sketch.html.

15. "... he ultimately developed many of the ideas that make the web ... but using paper cards instead of computers." A. Wright, "The secret history of hypertext," *The Atlantic*, May 22, 2014, www.theatlantic.com/technology/archive/2014/05/in-search-of-the-proto-memex/371385.

16. "Writing in 1892 ...": P. Otlet, Something about bibliography, ch. 1 in W. B. Rayward (ed. and translator), *International Organisation and Dissemination of Knowledge: Selected Essays of Paul Otlet*, Elsevier Science Publishers B.V., 1990, pp. 11–24, www.archive.org/stream/internationalorg00otle/internationalorg00otle_djvu.txt.

17. "... information customized to our needs at the moment of access." D. Berleant and H. Berghel, Customizing information: Part 1, *IEEE Computer*, Sept. 1994, vol. 27, no. 9, pp. 96–99, and Part 2, Oct. 1994, vol. 27, no. 10, pp. 76–78.

Chapter Five: *Live Anywhere, Work Anywhere Else*

1. "Commuting by car has been found to be a debilitating, continuous drain on our happiness and quality of life." Commuting, scienceblogs.com/cortex/2010/03/commuting.php.

2. "Some feel that urbanization is the future." R. Florida, *The Great Reset*, Harper, 2010.

3. "The World Wide Web is making us all part of a giant global digital village," H. Berghel, Digital Village, column in *Communications of the ACM*, 1995–2008, berghel.net/col-edit/ce.php.

4. "... content moderators sometimes known as data janitors." N. Resnikoff, Invisible data janitors mop up top web sites behind the scenes, *Al Jazeera America*, May 1, 2015, america.aljazeera.com/articles/2015/5/1/invisible-data-janitors-mop-up-top-websites-behind-the-scenes.html.

5. "Some telecommuters have been using internet-friendly establishments like Starbucks for some time already." K. Kauer, Starbucks: The new office space, Silicon Valley Mamas, Sept. 26, 2012, siliconvalleymamas.com/2012/09/starbucks-the-new-office-space.

Chapter Six: *From Highly Centralized to Highly Decentralized Societies*

1. "… which makes modern civilization so complex and interconnected that catastrophic disruptions may be both unpredictable and inevitable." C. Perrow, *Normal Accidents: Living with High-Risk Technologies*, Princeton University Press, 1999.

2. "No large refinery has been built in the US since the last one began operations in Garyville, Louisiana in 1977." (1) When was the last refinery built in the United States? U.S. Energy Information Administration, www.eia.gov/tools/faqs/faq.cfm?id=29&t=6. (2) Garyville Refinery: Garyville, Louisiana, www.marathonpetroleum.com/Operations/Refining_and_Marketing/Refining/Garyville_refinery.

3. "According to congressional testimony …": W. R. Graham, Chairman, Commission to Assess the Threat to the United States from Electromagnetic Pulse (EMP) Attack, Statement Before the House Committee on Homeland Security's Subcommittee on Emerging Threats, Cybersecurity, and Science and Technology, July 21, 2008, www.docstoc.com/docs/40760092/The-House-Committee-on-Homeland-Security-Subcommittee-on-Emerging.

4. "The electromagnetic pulse generated by a high altitude nuclear explosion is one of a small number of threats that can hold our society at risk of catastrophic consequences." (a) W. R. Graham (2008); (b) J. S. Foster, Jr., et al., Report of the Commission to Assess the Threat to the United States from Electromagnetic Pulse (EMP) Attack: Critical National Infrastructures, pub. by US Independent Agencies and Commissions, July 14, 2008, ISBN 9780160809279, www.empcommission.org/docs/A2473-EMP_Commission-7MB.pdf.

5. "some grid systems may experience complete collapse or blackouts …": NOAA Space Weather Scales, NOAA / Space Weather Prediction Center, www.swpc.noaa.gov/noaa-scales-explanation.

6. "… an especially perilous one …": T. Phillips, Near miss: The solar superstorm of July 2012, *Science@NASA Headline News*, July 23,

2014, science.nasa.gov/science-news/science-at-nasa/2014/23jul_superstorm.

7. "It could happen; there is no proof otherwise." C. Perrow, *Normal Accidents: Living with High-Risk Technologies*, Princeton University Press, 1999.

8. "A simple design strategy that can help is the *hardware interlock*." Second hand software: The Therac story, chapter 29 in T. A. Kletz, *Learning from Accidents*, 3rd ed., Gulf Professional Publishing, 2001.

9. "Rather, it was the end point of the path of least resistance, the macroeconomic result of innumerable microeconomic decisions, a rough shove from the iron fist of Adam Smith's invisible hand." (a) A. Smith, *The Theory of Moral Sentiments*, 1759, easily found on the web. (b) A. Smith, *The Wealth of Nations*, 1776, also easily found on the web.

10. "Constructing a nuclear reactor in your mother's shed is even less practical (yes, it's been tried)." K. Silverstein, *The Radioactive Boy Scout: The True Story of a Boy and his Backyard Nuclear Reactor*, Random House, 2004, etc. Or search the web for "David Hahn," the name of the boy scout.

11. "… greedy electric generation and distribution business." D. C. Johnston, Financiers and power producers rig electricity markets, *Al Jazeera America*, Feb. 27, 2015, america.aljazeera.com/opinions/2015/2/financiers-and-power-producers-rig-electricity-markets-rob-consumers.html.

12. "Solar paint containing biomolecular machinery extracted from ordinary plants as its 'secret ingredient' has already been invented," A. Mershin, et al., Self-assembled photosystem-I biophotovoltaics on nanostructured TiO_2 and ZnO, *Scientific Reports*, Feb. 2, 2012, vol. 2, art. no. 234, www.nature.com/srep/2012/120202/srep00234/full/srep00234.html.

13. "PV is generally not yet fully competitive with your local electric utility. But it's getting closer all the time." Photovoltaic Grid Parity Monitor (GPM), eclareon,www.eclareon.eu/en/gpm.

14. "... centralized electricity production in the US is down from a 2008 high." D. Gross, Beyond Energy Efficiency, *Slate*, July 30, 2014, www.webcitation.org/6ZcOpgudc.
15. "must-consider action" ... "directly threaten the centralized utility model." P. Kind, Disruptive challenges: financial implications and strategic responses to a changing retail electric business, Edison Electric Institute, 2013, www.eei.org/ourissues/finance/Documents/disruptivechallenges.pdf.
16. "**Figure 2. Shuttle vehicle with roof mounted solar panel.**" New Trojan Transit system rolls out March 30, *UALR Now*, ualr.edu/news/2015/03/25/new-trojan-transit-system-rolls-out-march-30.
17. "Kitchen food waste can be an excellent biogas source. Grass clippings are another source," Turning trash into power: Biological engineers generate natural gas with bacteria, *ScienceDaily*, Oct. 1, 2006, www.sciencedaily.com/videos/2006/1002-turning_trash_into_power.htm.
18. "Onsite Power Systems," onsitepowersystems.com.
19. "A technology for that is the microbial fuel cell (MFC)." (a) www.webcitation.org/6ZZ66J5gD; (b) G. Cook, New fuel cell uses germs to generate electricity, *Boston Globe*, Sept. 8, 2003, www.boston.com/news/local/articles/2003/09/08/new_fuel_cell_uses_germs_to_generate_electricity.
20. "In an MFC, bacteria live in a biofilm on the anode and decompose the waste, which is dissolved or in tiny particles suspended in water." D. Pant, et al., A review of the substrates used in microbial fuel cells (MFCs) for sustainable energy production, *Bioresource Technology*, Mar. 2010, vol. 101, issue 6, pp. 1533–1543.
21. "... urine-powered microbial fuel cells." Gates Foundation awards grants to develop urine powered fuel cells, waterless toilets and solar steam sterilizers, press release, Bill & Melinda Gates Foundation, www.gatesfoundation.org/Media-Center/Press-Releases/2013/12/Gates-Foundation-Awards-Grants-to-Waterless-Toilets.

22. "In very dry locales, energy efficient water recyclers can work, such as the 'slingshot,' " www.youtube.com/watch?v=s_8DEDctH5I, etc.

23. "Prototype tomatobots were invented by 2009," J. Lowe, Robots tend the tomatoes, *Christian Science Monitor*, Mar. 11, 2009, www.csmonitor.com/The-Culture/Gardening/diggin-it/2009/0311/robots-tend-the-tomatoes.

24. "It was named IAM-BRAIN because it was built by the Institute of Agricultural Machinery's Bio-oriented Technology Research Advancement Institution." T. Hornyak, Strawberry-picking robot knows when they're ripe, CNET News, Dec. 13, 2010, news.cnet.com/8301-17938_105-20025402-1.html.

25. "Veggiebot commercialization got a boost in 2012 when Blue River Technology announced $3.1 million in funding to develop agribots based on its prototype Lettuce Bot ...": T. Hornyak, Down on the farm, Lettuce Bot is quietly destroying weeds, *CNET News*, Sept. 14, 2012, news.cnet.com/8301-17938_105-57513147-1/down-on-the-farm-lettuce-bot-is-quietly-slaying-weeds.

Chapter Seven: *When Genomes Get Cheap*

1. "The US government spent $2.7 billion ...": The human genome project completion: Frequently asked questions, National Human Genome Research Institute, www.genome.gov/11006943.

2. "James Watson ...": E. Singer, The $2 Million Genome, *Technology Review*, June 1, 2007, www.technologyreview.com/Biotech/18809.

3. "[w]ith a steep drop in the costs of sequencing and an explosion of research on genes, medical experts expect that genetic analyses of cancers will become routine." G. Kolata, In treatment for leukemia, glimpses of the future, *New York Times*, July 8, 2012, www.nytimes.com/2012/07/08/health/in-gene-sequencing-treatment-for-leukemia-glimpses-of-the-future.html.

4. "Officially, an orphan disease ...": Rare diseases and related terms, National Institutes of Health Office of Rare Diseases, rarediseases.info.nih.gov/RareDiseaseList.aspx?PageID=1.

5. "The human version of the FOXP2 gene was among the first to be identified as necessary for our language abilities." S. E. Fisher and C. Scharff, FOXP2 as a molecular window into speech and language, *Trends in Genetics*, 2009, vol. 25, no. 4, pp. 166–177.

6. "… 'CRISPR'-based genome editing was used to edit the DNA of mouse sperm cells in 2015." L. Cannon, CRISPR prevents cataracts, *leahcanscience.com*, February 9, 2015, leahcanscience.com/2015/02/09/crispr-prevents-cataracts.

7. "CRISPR is so powerful that in March, 2015 scientists proposed banning its use on humans." N. Wade, Scientists seek ban on method of editing the human genome, *New York Times*, www.nytimes.com/2015/03/20/science/biologists-call-for-halt-to-gene-editing-technique-in-humans.html

8. "… scientists in China experimentally genetically modified fertilized human eggs." A. Regalado, Chinese team reports gene-editing human embryos, *Technology Review*, April 22, 2015, www.technologyreview.com/news/536971/chinese-team-reports-gene-editing-human-embryo.

9. "Actual genetically modified people could be as little as 10 years away." A. Regalado, Engineering the perfect baby, *Technology Review*, March 5, 2015, www.technologyreview.com/featuredstory/535661/engineering-the-perfect-baby.

10. "Ultimately, superbabies will be designed and raised …": See the movie *Gattaca*, 1997, starring Ethan Hawke, Uma Thurman, and Jude Law.

11. "Already, a technique called PGD, for preimplantation genetic diagnosis, is being used for gender selection." J. Sidhu, How to buy a daughter, *Slate*, Sept. 14, 2012, www.slate.com/articles/health_and_science/medical_examiner/2012/09/sex_selection_in_babies_through_pgd_americans_are_paying_to_have_daughters_rather_than_sons_.html.

Chapter Eight: **Cheaper Teaching, Faster Learning**

1. "… when the first records of classes taught by an instructor appear.": E. Strouhal, *Life of the Ancient Egyptians*, University of Oklahoma Press, 1992, p. 36, ISBN 080612475X.
2. "The tomb of Kheti … calls upon graduates to behave well." M. Parsons, Education in Ancient Egypt, www.touregypt.net/featurestories/educate.htm.
3. "… Stanford University offered 'Introduction to Artificial Intelligence' online in 2011 to 160,000 students … the professor founded startup company Udacity …": www.udacity.com. Stanford also offered a machine learning class the same semester: C. Wilson, Blogging the Stanford machine learning class, *Slate*, Dec. 20, 2011, www.slate.com/articles/technology/future_tense/features/2011/learning _machine/machine_learning_to_predict_presidential_elections_.html.
4. "'It's pretty obvious that degrees will go away' since 'this model isn't valid anymore.'" G. Anders, How would you like a graduate degree for $100, *Forbes Magazine*, June 25, 2012, www.forbes.com/forbes/2012/0625/technology-sebastian-thrun-udacity-university-of-disruption.html.
5. "… he now saw Udacity as becoming the 'Ikea' of education. J. Moules, Udacity founder takes on virtual study revolution, *Financial Times*, July 26, 2015, www.webcitation.org/6aaWzQbQO.
6. "The three must have started out seeing eye to eye, but ended up cross-eyed." A. Watters, Stanford professors Daphne Koller & Andrew Ng also launching a massive online learning startup, *Hack Education blog*, Jan. 31, 2012, www.hackeducation.com/2012/01/31/stanford-professors-daphne-koller-and-andrew-ng-launch-coursera.
7. "… in Summer 2013 Stanford itself adopted a 3rd competing MOOC, OpenEdX." *Stanford Report*, June 11, 2013, news.stanford.edu/news/2013/june/open-source-platform-061113.html.
8. "One might think an ideal topic for a huge online course would be how to teach huge online courses …": V. Strauss, How online class about online learning failed miserably, *Washington Post*, Feb. 5, 2013,

www.washingtonpost.com/blogs/answer-sheet/wp/2013/02/05/how-online-class-about-online-learning-failed-miserably.

9. "Let us expand that concept by also teaching people to identify the statement writer's *agenda* and thus any incentive to manipulate the reader's mind." A. Applebaum, The trolls among us, *The Atlantic*, November 28, 2014, www.slate.com/articles/news_and_politics/foreigners/2014/11/internet _trolls_pose_a_threat_internet_commentators_shouldn_t_be_anonymo us.html.

10. "Call them eartop computers." M. Cohen, J. Herder, and W. L. Martens, Cyberspacial Audio Technology, *Journal of the Acoustical Society of Japan (E)*, Nov. 1999, vol. 20, no. 6, web-ext.u-aizu.ac.jp/~mcohen/welcome/publications/JASJ-reviewE.pdf.

11. "... author Larry Niven's 'wireheads' ...": In some of his sci-fi works, wireheads were persons addicted to electrical stimulation of the brain's pleasure center.

Chapter Nine: ***Soylent Spring***

1. "But the earliest may have been New Guinea, with intriguing evidence suggesting taro and yam farming over 12,000 years ago. The nearby Solomon Islands may have preceded even that, with some evidence of taro cultivation well over 20,000 years ago." R. Fullagar, J. Field, T. Denham & C. Lentfer, Early and mid-holocene tool-use and processing of taro (Colocasia esculenta), yam (Discorea sp.) and other plants at Kuk Swamp in the highlands of Papua New Guinea, *Journal of Archaeological Science*, 2006, vol. 33, no. 5, pp. 595–614.

2. "... like the small quantities of arsenic often fed (deliberately!) to chickens." N. D. Kristof, Arsenic in our chicken? *New York Times*, Apr. 4, 2012, www.nytimes.com/2012/04/05/opinion/kristof-arsenic-in-our-chicken.html.

3. "... cattle fart into the atmosphere about 80 million tons (73 million tonnes) per year of the potent greenhouse gas methane ...": Methane, EPA, www.epa.gov/methane/sources.html.

4. "PETA (People for the Ethical Treatment of Animals) offered a $1 million prize in 2008 for a commercially viable process." PETA offers $1 million reward to first to make in vitro meat, www.peta.org/features/In-Vitro-Meat-Contest.aspx.

5. "… on August 5, 2013, the first hamburger made from vat-grown meat was fried up and gulped down, at the bargain basement price of only 250,000 Euros (about 330,000 US dollars)." A. Madrigal, Chart: When will we eat burgers grown in test tubes? *The Atlantic*, Aug. 6, 2013, www.theatlantic.com/technology/archive/2013/08/chart-when-will-we-eat-hamburgers-grown-in-test-tubes/278405. Also see: Cultured beef, Maastricht University, culturedbeef.net.

6. *The In Vitro Meat Cookbook*, Next Nature, 2014, ISBN 978-9063693589, www.webcitation.org/6TSIOyC9r.

7. "One such product that came on the market in 2012 was, according to one reviewer, 'so good it will freak you out.'" F. Manjoo, Fake meat so good it will freak you out, *Slate*, July 26, 2012, www.slate.com/articles/technology/technology/2012/07/beyond_meat_fake_chicken_that_tastes_so_real_it_will_freak_you_out_.html.

8. "… was recognized as such as early as the 1890s," H. J. W. Dam, Foods in the year 2000. Professor Berthelot's theory that chemistry will displace agriculture. *McClure's Magazine*, Sept. 1894, pp. 303–312, www.unz.org/Pub/McClures-1894sep-00303. See also maureenogle.com/2013/08/06.

Chapter Ten: *The Turbulence of Short-Term Action*

1. *The Ant and the Grasshopper*: Aesop's Fables.

2. "… each cigarette shortens life by an average of 11 minutes …": A web search on terms like '11,' 'minutes,' 'cigarettes' and 'life' returns many links to relevant information.

3. "Adolescents can act on impulse as though they think they are immortal." V. F. Reyna and F. Farley, Risk and Rationality in Adolescent Decision Making: Implications for Theory, Practice, and Public Policy,

Psychological Science in the Public Interest, Sept. 2006, vol. 7, no. 1, www.psychologicalscience.org/pdf/pspi/pspi7_1.pdf.

4. "Adults often think that souls really make them immortal ...": Billions of adult adherents of various major religions believe that their sentient essences are, indeed, immortal. See also J. Bering, Never say die: Why we can't imagine death ... why so many of us think our minds continue on after we die, *Scientific American*, Oct. 2008, www.sciam.com/article.cfm?id=never-say-die.

5. "The former Easter Island civilization experienced destruction of all its palm trees ...": T. L. Hunt and C. P. Lipo, Ecological catastrophe, collapse, and the myth of 'ecocide' on Rapa Nui (Easter Island), chapter 2 in P. A. McAnany and N. Yoffee, *Questioning Collapse*, 2010, Cambridge University Press.

6. "The 7 habits of highly effective people": S. Covey, *The 7 Habits of Highly Effective People*, 1990, Free Press.

7. "Begin with the end in mind." S. Covey (1990).

8. "Eat dessert first." From quote "Life is uncertain. Eat dessert first." Attributed to E. Ulmer, thinkexist.com/quotation/life-is-uncertain-eat-dessert-first/347441.html.

9. "Teaching this skill would thus be a worthwhile goal." W. Mischel, Y. Shoda, and M. L. Rodriguez, Delay of gratification in children, chapter 15 in *Social Psychology: A General Reader*, edited by A. W. Kruglanski and E. T. Higgins, Psychology Press, 2003, pp. 202–211.

10. "David Collingridge identified an important difficulty," summary in E. Morozov, The Collingridge dilemma, *Edge*, 2012, edge.org/response-detail/10898. More details in: D. Collingridge, *The Social Control of Technology*, Open University Press, 1981.

11. "The value of being driven by long-term considerations is responsible in part for the success of Amazon, for example." M. Yglesias, Amazon is a great company because it has the most generous shareholders in the world, *Slate*, Dec. 12, 2012, www.slate.com/blogs/moneybox/2012/12/12/amazon_s_zero_profit_bu siness_strategy_it_s_amazing_but_someday_we_may_all.html.

Chapter Eleven: **Battle for the Mind**

1. "... the title of a book by W. Sargant ...": *Battle for the Mind, a Physiology of Conversion and Brainwashing*, Doubleday & Company, 1957; Malor, 1997.
2. "In 1988, shocking memories of childhood sexual and even satanic ritual abuse began to sweep America." R. Ofshe and E. Watters, *Making Monsters: False Memory, Satanic Cult Abuse, and Sexual Hysteria*, University of California Press, 1996. See also False Memory Syndrome Foundation, fmsonline.org.
3. "Many of these recovered memories were false – despite the serious reality of childhood sexual and other abuse." E.g. M. Fontaine, America has an incest problem, *The Atlantic*, Jan. 24, 2013, www.theatlantic.com/national/archive/2013/01/america-has-an-incest-problem/272459.
4. "Loftus's early research looked at the malleability of memories of vehicular collisions." E. F. Loftus and J. C. Palmer, Reconstruction of automobile destruction: An example of the interaction between language and memory, *Journal of Verbal Learning and Verbal Behavior*, 1974, vol. 13, pp. 585–589, webfiles.uci.edu/eloftus/LoftusPalmer74.pdf.
5. "In one test, a third of eye witnesses remembered an experimental 'perpetrator' from a lineup that in fact did not include the perpetrator." W. Saletan, Leading the Witness, *Slate*, May 26, 2010, www.slate.com/id/2251882/pagenum/all.
6. "... persuaded up to 18% of people that an implausible event — witnessing demonic possession as a child — probably happened to them personally." G. Mazzoni, E. Loftus, and I. Kirsch, Changing beliefs about implausible autobiographical events: A little plausibility goes a long way, *Journal of Experimental Psychology: Applied*, 2001, vol. 7, no. 1, pp. 51–59.
7. "Yet in a later experiment, 16% remembered shaking hands with Bugs Bunny at Disneyland." K. Braun, R. Ellis, and E. Loftus, Make my memory: How advertising can change our memories of the past,

Psychology & Marketing, Jan. 2002, vol. 19, no. 1, pp. 1–23, faculty.washington.edu/eloftus/Articles/BraunPsychMarket02.pdf.

8. "The power of this technique relies on the well-known capability of social pressure to modify memory." M. B. Reysen, The effects of social pressure on false memories, *Memory & Cognition*, 2007, vol. 35, no. 1, pp. 59–65.

9. "to accept a false reality as truth … is the very essence of madness." R. Frederickson, *Repressed Memories*, Simon & Schuster, 1992, p. 160.

10. "Hallucinations, whether revelatory or banal, are not of supernatural origin; they are part of the normal range of human consciousness and experience." O. Sacks, Seeing God in the third millennium, *The Atlantic*, Dec. 12, 2012, www.theatlantic.com/health/archive/2012/12/seeing-god-in-the-third-millenium/266134.

11. "relatives confirmed that her aunt … had found the body." W. Saletan, Leading the Witness, *Slate*, May 26, 2010, www.slate.com/id/2251882/pagenum/all.

12. *We can remember it for you wholesale*: P. K. Dick, *The Magazine of Fantasy & Science Fiction*, Apr. 1966, and many collections.

13. "Will psychotherapists do it for our own good, for example implanting memories of getting sick from ice cream to improve our nutrition and reduce obesity, as proved possible by Loftus?" D. M. Bernstein, C. Laney, E. K. Morris, and E. F. Loftus, False beliefs about fattening foods can have healthy consequences, *Proceedings of the National Academy of Sciences* (*PNAS*), Sept. 27, 2005, vol. 102, no. 39, pp. 13724–13731, www.pnas.org/content/102/39/13724.full.pdf.

14. "Over the next 50 years we will further master the ability to create false memories." … "The most potent recipes may involve pharmaceuticals that we are on the brink of discovering." Elizabeth Loftus forecasts the future, *New Scientist*, Nov. 18, 2006, issue 2578, www.newscientist.com/article/mg19225780.112.

15. "A concerted set of messages aimed at influencing the opinions of behavior of large numbers of people." See en.wiktionary.org/wiki/propaganda.

16. "To help guard and protect memory, the 'insidious mechanisms and methods' of memory modification should be taught ...": K. A. Braun and E. F. Loftus, Advertising's misinformation effect, *Applied Cognitive Psychology*, 1998, vol. 12, pp. 569–591, webfiles.uci.edu/eloftus/BraunLoftusAdvertisingMisinfoACP98.pdf.

17. "For example, people should know that memories from adolescence are the best targets for beer advertisers." K. A. Braun, R. Ellis, and E. F. Loftus, Make my memory: How advertising can change our memories of the past, *Psychology & Marketing*, Jan. 2002, vol. 19, no. 1, pp. 1–23, faculty.washington.edu/eloftus/Articles/BraunPsychMarket02.pdf.

18. "People should also know that memory revisionists 'want the consumer to be involved enough [to] process the false information,' yet not enough to 'notice the discrepancy between the advertising information and their own experience.'" K. A. Braun-LaTour, M. S. LaTour, J. E. Pickrell, and E. F. Loftus, How and when advertising can influence memory for consumer experience, *Journal of Advertising*, winter 2004, vol. 33, no. 4, pp. 7–25, webfiles.uci.edu/eloftus/BraunLaTourPickLoftusJofAd04.pdf.

19. ... "imagining a childhood event inflates confidence that it occurred." M. Garry, C. G. Manning, and E. F. Loftus, Imagining inflation: Imagining a childhood event inflates confidence that it occurred, *Psychonomic Bulletin & Review*, 1996, vol. 3, no. 2, pp. 208–214.

20. "This imagining technique can be used to 'recall' memories from infancy, before birth, and even as a sperm or egg cell!" (1) J. Sadgar, Preliminary study of the psychic life of the fetus and the primary germ, *Psychoanalytic Review*, July 1941, vol. 28, no. 3, pp. 327–358. (2) L. R. Hubbard, *Dianetics: The Modern Science of Mental Health*, 1950.

21. "And if that is not early enough, the same method suffices to recall 'memories' from past lives." International Association of Past Life Therapists, www.pastlives.net.

22. "Memory, like liberty, is a fragile thing." E. F. Loftus and W. H. Calvin, Memory's future, *Psychology Today*, Mar.–Apr. 2001, vol. 34, no. 2, pp. 55+.

Chapter Twelve: *Will Artificial Intelligence Threaten Civilization?*

1. "*... protest march against robots.*" Stop the robots, March 2015, www.webcitation.org/6X6MVJWRT.

2. "*... the Austin, Texas demonstration later turned out to be nothing more than a marketing stunt ...*": C. J. Anders, Yep, that anti-robot protest at SXSW was a marketing stunt [updated], *io9,* March 16, 2015, io9.com/yeah-that-stop-the-robots-protest-group-is-probably-1691759434.

3. "*...* computer performance on a test of artificial intelligence, the classic 'Turing Test,' has been increasing over the years, with chatbot competitions held every year since 1991." Loebner prize competition, www.loebner.net/Prizef/loebner-prize.html.

4. "As noted first by Asimov in 1956 and later by Good, Vinge, Moravec, Kurzweil, and others," (1) Isaac Asimov wrote, "Each [intelligent computer] designed and constructed its ... more intricate, more capable successor" in: The last question, *Science Fiction Quarterly*, Nov. 1956. Available on the web. (2) Irving J. Good, Speculations concerning the first ultraintelligent machine, in F. L. Alt and M. Rubinoff, editors, *Advances in Computers*, 1965, vol. 6, pp. 31–88. Available on the web. (3) V. Vinge, The coming technological singularity: How to survive in the post-human era, NASA technical report CP-10129, 1993, and *Whole Earth Review*, winter, 1993. Available on the web. (4) R. Kurzweil, *The Singularity is Near*, 2005.

5. "*...* a robotic closet that '... a woman may step into ... which will ... put on clothes most suited to the occasion and fix her hair ...'": J. McCarthy, Robot servants, 1995 (updated 2002), www-formal.stanford.edu/jmc/future/robots.html.

6. "Shades of the classic sci-fi stories ...": See "Robbie," the first story in Asimov's short story collection *I, Robot*, 1950 (originally "Strange Playfellow," *Super Science Stories* magazine, Sept. 1940). See also "Nanny," 1955, reprinted in *The Book of Philip K. Dick*, 1973.

7. "intellects vast and cool and unsympathetic" H. G. Wells, *The War of the Worlds*, 1898, www.gutenberg.org/ebooks/36.

8. "… invisible hand …": Phrase coined by famed economist Adam Smith, *The Wealth of Nations*, first published in 1776, www.gutenberg.org/etext/3300.

9. "… probably, most animal species are parasites." R. Nuwer, Parasitism is the most popular lifestyle on Earth, *New Scientist*, Issue 2927, July 29, 2013, www.webcitation.org/6Zbx9Lszn.

10. "… parasites often become extremely simplified …": S. J. Gould, *The Structure of Evolutionary Theory*, Harvard University Press, 2002, p. 1150.

11. "The movement to ban AI-enabled weaponry finally began in earnest with an open letter to the public on July 28, 2015." Autonomous weapons: An open letter from AI & robotics researchers, Future of Life Institute, 2015, futureoflife.org/AI/open_letter_autonomous_weapons.

12. "… genie bearing wishes …": *One Thousand and One Nights*, many editions and publishers.

13. "… Frankenstein …": M. Shelley, *Frankenstein; or, the Modern Prometheus*, www.literature.org/authors/shelley-mary/frankenstein/index.html.

14. "… three laws of robotics …": I. Asimov, *I, Robot* (the book, not the movie, which bears little resemblance to the book). Also see R. Clarke, Asimov's Laws of Robotics: Implications for information technology, part 1, *IEEE Computer*, Dec. 1993, vol. 26, no. 12, pp. 53–61, and part 2, Jan. 1994, vol. 27, no. 1, pp. 57–66.

Chapter Thirteen: *Deconstructing Nuclear Nonproliferation*

1. "… Stanislav Petrov was honored … for single-handedly heading off nuclear war between the United States and the Soviet Union …": P. Aksenov, Stanislav Petrov: The man who may have saved the world, *BBC News*, September 26, 2013, www.bbc.com/news/world-europe-24280831.

2. "Let's examine this a little more closely." E.g. Cold war and proliferation, Institute for Structure and Nuclear Astrophysics, University of Notre

Dame,
www.nd.edu/~nsl/Lectures/phys205/pdf/Nuclear_warfare_5.pdf.

3. "It is difficult to predict, especially about the future," quote originating in Denamark and variously attributed, quoteinvestigator.com/2013/10/20/no-predict.

4. "... Ukraine ... Kazakhstan ... Belurus ... finally persuaded to do so." Nuclear Forces Guide, Federation of American Scientists, www.fas.org/nuke/guide.

5. "His response: The ideas in new writings were simply repeating those in existing ones, and what is the point of more research when it would just be redundant?" I saw it on TV during his period in office and, shocked, still remember it.

Chapter Fourteen: *Space Empire — From Mercury to Neptune and Beyond*

1. "The solar system is ... a pretty soggy place." M. Kramer, Jupiter's moon Ganymede has a salty ocean with more water than Earth, *Space.com*, March 12, 2015, www.space.com/28807-jupiter-moon-ganymede-salty-ocean.html.

2. "Water exists in large quantities on Mercury in the form of ice, ...": MESSENGER finds new evidence for water ice at Mercury's Poles, NASA press release, Nov. 29, 1012, www.nasa.gov/mission_pages/messenger/media/PressConf20121129.ht ml.

3. "... a day-night cycle on Mercury lasts two Mercurian years, or 176 Earth days ...": See the animated simulator at www.sciencenetlinks.com/interactives/messenger/dom/DayOnMercury. html.

4. "A 2-mile (3 km) diameter balloon containing a breathable atmosphere ... would generate millions of tons of lift in the denser, carbon dioxide-rich Venusian atmosphere at habitable altitude." G. A. Landis, Colonization of Venus, *Proceedings of the Space Technology and Applications International Forum*, Albuquerque, Feb. 2–6, 2003,

5.pdf. (Update: A 2014 proposal is at goo.gl/nmR1Xx.)

5. "One approach to finding a place to hang your hat on the Moon is mooncaves …": A readable survey is: I. O'Neill, Living in lunar lava tubes, *Discover News*, Oct. 27, 2009, news.discovery.com/space/moon-lunar-lava-skylight.html.

6. "The best currently known location for Lunaria is the Mount Yewridge area near the South Pole." A number of images are at images.google.com/images?q="peak+of+eternal+light". Also see M. Kruijff, The peaks of eternal light on the lunar South Pole: How they were found and what they look like, in the *4th International Conference on Exploration and Utilization of the Moon* (*ICEUM4*), ESA/ESTEC, SP-462, Sept. 2000, www.academia.edu/1215974/THE_PEAKS_OF_ETERNAL_LIGHT_ON_THE_LUNAR_SOUTH_POLE_How_they_were_found_and_what_they_look_like.

7. "Even liquid water has been found on Mars!" NASA confirms evidence that liquid water flows on today's Mars, *JPL News*, Sept. 28, 2015, www.jpl.nasa.gov/news/news.php?feature=4722.

8. "One is the knowledge of Earth's long cultural fascination with Mars." R. Crossley, *Imagining Mars*, Wesleyan University Press, 2011.

9. "Geometry tells us that to a person standing on the surface, Earth's horizon is only a little over twice as far as Ceres'." This is based on the Pythagorean theorem applied to a right triangle where the side of interest spans from the observer's eye to the horizon (the point where her gaze fondly skims the surface of the sphere), another side is a radius line ending at that horizon point, and the hypotenuse is a line straight from the center of the planetary sphere to her eye.

10. "This would propel the youths forward at that speed …": An 8-foot pole used by an individual to extend 2 feet of arm length gives a radius of 10 feet, diameter of 20 feet, and circumference of approximately 63 feet. Thus swinging a pole full circle at the leisurely rate of every 2 seconds gives a forward speed of about 31.5 ft./sec., or 21 miles (34 km) per hour.

11. "Approximate time to begin meditative wakeful state (was called 'watching' on Earth before artificial lighting became common)." See T. A. Wehr, D. E. Moul, G. Barbato, H. A. Giesen, J. A. Seidel, C. Barker, and C. Bender, Conservation of photoperiod-responsive mechanisms in humans, *American Journal of Physiology — Regulatory, Integrative and Comparative Physiology*, 1993, vol. 265, issue 4, pp. R846–R857. Summarized in A. Aubrey, How aging changes sleep patterns, Aug. 3, 2009, *NPR*, www.npr.org/templates/story/story.php?storyId=111415462.

12. "Bullets could be fired into orbit." See e.g. answers.yahoo.com/question/index?qid=20090813164440AAw9JGz for sample calculations.

13. "On the other hand, gas giants are not without temptation: Some scientists suggest that far enough down, large diamond bergs drift in liquid carbon, which might tempt some to try to fish them out." J. H. Eggert, et al., Melting temperature of diamond at ultrahigh pressure, *Nature Physics*, vol. 6, pp. 40–43 (2010), www.nature.com/nphys/journal/v6/n1/abs/nphys1438.html.

14. "Predicted to lurk far, far away, … if it really exists." K. Fesenmaier, Caltech researchers find evidence of a real ninth planet, *Now@Caltech*, Jan. 20, 2016, www.caltech.edu/news/caltech-researchers-find-evidence-real-ninth-planet-49523.

15. "… astronomers have reported a planet 4,000 light years away that is likely to be mostly a giant diamond." M. Bailes, et al., Transformation of a Star into a Planet in a Millisecond Pulsar Binary, Science, Aug. 25, 2011, www.sciencemag.org/content/early/2011/08/19/science.1208890.full.pdf, also see arxiv.org/PS_cache/arxiv/pdf/1108/1108.5201v1.pdf.

16. "Rather than mere ordinary liquid rock, however, large quantities of sulfur issue forth." C. J. Hamilton, Io (www.solarviews.com/eng/io.htm), in Views of the Solar System, a website. Actually, sulfur may issue from volcanoes here on Earth more often than is commonly realized (see A. J. L. Harris, S. B. Sherman, and R. Wright, Discovery of self-combusting volcanic sulfur flows,

Geology, vol. 28, no. 5, pp. 415–418, May 2000,
www.higp.hawaii.edu/~wright/geology28.pdf.).

17. "That's actually plenty of light to see around in, equivalent to shortly
after sunset on Earth." Pluto time, *NASA*, update of June 5, 2015,
solarsystem.nasa.gov/plutotime.

18. "…synthesizing nutrients artificially in a chemical laboratory, an idea
that follows the circa 1894 proposals of French chemist Marcellin
Berthelot." H. J. W. Dam, Foods in the year 2000. Professor Berthelot's
theory that chemistry will displace agriculture. *McClure's Magazine*,
Sept. 1894, pp. 303–312, www.unz.org/Pub/McClures-1894sep-00303.
See also www.maureenogle.com/maureen-ogle/2013/08/06/meat-from-
the-laboratory-back-to-the-future-with-a-bit-of-history.

19. "In fact, there might be 100,000 times as many such nomads floating
around in space as there are stars." L. E. Strigari, M. Barnabe, P. J.
Marshall, and R. D. Blandford, Nomads of the Galaxy, draft paper,
2012, arxiv.org/pdf/1201.2687v1.pdf.

20. "… the US National Ignition Facility reported getting more energy from
fusion than was needed as input to make the fusion happen." Scientists
achieve fuel gain exceeding unity in confined fusion implosion,
Kurzweil Accelerating Intelligence Newsletter, February 13, 2014,
www.kurzweilai.net/scientists-achieve-fuel-gain-exceeding-unity-in-
confined-fusion-implosion.

21. "… an estimated 40 billion such planets in the Milky Way galaxy." S.
Borenstein, Study: 8.8 billion Earth-size, just-right planets, Associated
Press, Nov. 4, 2013, bigstory.ap.org/article/study-88-billion-earth-
sized-just-right-planets.

Chapter Fifteen: *Tastes Like the Singularity*

1. "not only queerer than we suppose, but queerer than we *can* suppose." Last paragraph of J. B. S. Haldane's essay Possible Worlds, e.g. in *Possible Worlds and Other Papers*, Harper and Brothers Publishers, 1928, p. 298.

2. "… money on line 53 of the infamous Connecticut 2008 income tax form CT-1040 …": See for example www.ct.gov/drs/lib/drs/forms/2008forms/incometax/ct-1040.pdf. If your modified Connecticut adjusted gross income was zero in that year, you would have hit a singularity had you chosen to try to fill out the form.

3. "All your base are belong to us." A passage in the English translation of the Japanese video game Zero Wing that went viral in the early 21st century. See e.g. www.youtube.com/watch?v=rfMC2aVhYuo.

4. "… Turing Test, created by British code breaker and war hero Alan Turing." A. M. Turing, Computing machinery and intelligence, *Mind*, Oct. 1950, vol. LIX, no. 236, pp. 433–460. E.g. mind.oxfordjournals.org/cgi/reprint/LIX/236/433.

5. ELIZA: J. Weizenbaum, Contextual understanding by computers, *Communications of the ACM*, Aug. 1967, vol. 10, no. 8, pp. 474–480. See also J. Weizenbaum, *Computer Power and Human Reason*, W. H. Freeman and Co., 1976.

6. "ELIZA created the most remarkable illusion of having 'understood' in the minds of the many people who conversed with it." J. Weizenbaum (1976), p. 189.

7. "Loebner Prize": See www.loebner.net/Prizef/loebner-prize.html.

8. "As a side note …": See loebner.net/Prizef/minsky.html.

9. "One measures a computer's creativity …": Georgia Tech professor proposes alternative to 'Turing Test,' *Georgia Institute of Technology*, November 19, 2014, www.webcitation.org/6UM7FELHE.

10. "Progressively increasing computer chess performance …": E.g. R. Kurzweil, *The Singularity is Near*, Penguin Books, 2005, pp. 274–278.

11. "By the year 2050, develop a team of fully autonomous humanoid robots that can win against the human world soccer champion team." www.webcitation.org/6ZZAKKV6u; originally accessed via www.robocup.org.

12. *I, pencil*: L. E. Read, *The Freeman*, Dec. 1958.

13. "... human societies of thousands, once isolated, have lost even basic pre-industrial technologies." G. Clark, *A Farewell to Alms: A Brief Economic History of the World*, Princeton University Press, 2007.

Chapter Sixteen: *Chasing the Future — Spoilsports of the Prediction Game*

1. "The social competition theory of human brain evolution even holds that game theory is the key to why we are intelligent." D. H. Bailey and D. C. Geary, Hominid brain evolution: Testing climatic, ecological, and social competition models, *Human Nature*, 2009, vol. 20, no. 1, pp. 67–79.

2. "The probability of being tipped enough to lose balance is small enough that a single such pencil would be unlikely to fall for a long time." D. Easton, The quantum mechanical tipping pencil — a caution for physics teachers, *European Journal of Physics*, 2007, vol. 28, pp. 1097–1104 (see p. 1103).

3. "Existential nihilism — the distressing feeling 'that the world lacks meaning or purpose' ...": E.g. www.allaboutphilosophy.org/existential-nihilism-faq.htm.

4. "Eat dessert first." Quote attributed to Ernestine Ulmer.

5. "Eat, drink and be merry, for tomorrow we shall die." *Isaiah* 22:13.

6. "Don't worry, be happy": Meher Baba, 1930s, description at www.avatarmeherbaba.org/erics/glossc-d.html. Borrowed as title of Grammy award-winning song by Bobby McFerrin, 1988.

7. "In fact, that is essentially what people do. Business decisions focus on short-term payback, with 'long-term planning' designating horizons as short as 3 years out." E.g. www.webcitation.org/6ZZB4I77y.

8. "societies choose to fail": J. Diamond, *Collapse: How Societies Choose to Fail or Succeed*, Penguin Group, 2004.

9. "If I am not for myself, who will be for me? And when I am only for myself, what am I? And if not now, when?" R. Hillel, *Pirkei Avot* 1:14.

10. "time value of money": TVM is standard terminology in the finance and accounting world.

11. "It does but, it turns out, not very much." There is a formula for calculating the sum of a geometrically decreasing, infinite series. Look it up (or even better, play with a spreadsheet instead).

Chapter Seventeen: **Warm, Poison Planet**

1. There is a fair amount of both popular and scientific literature on this topic. A well-known full length work that helps bridge the gap between those literatures is: P. D. Ward, *Under a Green Sky*, HarperCollins, 2007.

2. "... eruptions released nickel, which nourished a bacterial bloom of Methanosarcina bacteria." C. Wald, Archaeageddon: how gas-belching microbes could have caused mass extinction, *Nature News*, Mar 31, 2014, www.nature.com/news/archaeageddon-how-gas-belching-microbes-could-have-caused-mass-extinction-1.14958.

Chapter Eighteen: **Day of Contact**

1. "... gravitational wormhole travel ...": This may be possible as it does not violate the laws of physics as presently known, according to M. Kaku, *Physics of the Impossible*, Random House, Inc., 2008, p. XVII & ch. 11.

2. "... chill sneer of command": Paraphrased from Percy Shelley's poem *Ozymandias*, 1818, www.online-literature.com/shelley_percy/672.

Chapter Nineteen: *"Darwin, Meet God."*
"Pleased to Meet You."

1. *Omphalos: An Attempt to Untie the Geological Knot*, by P. Gosse, 1857, www.gutenberg.org/ebooks/39910.

2. "Never was a book cast upon the waters with greater anticipations of success ...": E. Gosse, *Father and Son: A Study of Two Temperaments*, 1907, www.gutenberg.org/ebooks/2540.

3. *Fads and Fallacies in the Name of Science*, by M. Gardner, Dover Publications, 1957, pp. 123-124.

4. "But what is so desperately wrong about omphalos?" S. J. Gould, *Adam's Navel*, Penguin Books, 1995.

5. "God might have created ... the world with all the marks of antiquity and completeness which it now exhibits." F.-R. de Chateaubriand, *Génie du Christianisme*, 1802, catalog.hathitrust.org/api/volumes/oclc/5745473.html.

6. "...Descarte...": R. Descartes, *Meditations on First Philosophy*, 1641 (in Latin), English translations currently available, such as at oregonstate.edu/instruct/phl302/texts/descartes/meditations/meditations.html.

7. "Now I do not know whether I was then a man dreaming I was a butterfly, or whether I am now a butterfly dreaming I am a man." Z. Zhuang, Zhuang Zhu dreamed he was a butterfly, in Chapter 2 of *Zhuangzi*, English translations currently available, such as at www.the-philosopher.co.uk/butter.htm.

8. "Logically, it's a genuine possibility." A. Marchand, *The Universe Is a Dream: The Secrets of Existence Revealed*, Inspired Arts Press, 2011.

9. "Singularitarians seriously want to upload their minds into computers as a way to mentally survive the deaths of their bodies." K. Wiley, *A Taxonomy of Mind Uploading*, Alautun Press, 2014, www.amazon.com/Taxonomy-Metaphysics-Mind-Uploading-Keith-Wiley/dp/0692279849. Also see *International Journal of Machine Consciousness* (Special Issue on Mind Uploading), June, 2012, vol. 4, no. 1.

10. "facing a genuine extinction of beingness — a real death." J. Bering, Why we can't imagine death, pragmasynesi.wordpress.com/2008/10/01/why-we-cant-imagine-death.

11. "Might we live in a simulated universe?" N. Bostrom, Are you living in a computer simulation?, *Philosophical Quarterly*, 2003, vol. 53, no. 211, pp. 243-255, www.simulation-argument.com.

12. "… even reincarnation," I. Stevenson, *Twenty Cases Suggestive of Reincarnation*, University of Virginia Press, 1980.

Chapter Twenty: *In Memory of Daylight Savings Time, R.I.P.*

1. "First proposed in 1885 by postal worker and entomologist George V. Hudson …": G. V. Hudson, On seasonal time-adjustment in countries south of lat. 30°, *Transactions and Proceedings of the Royal Society of New Zealand*, vol. 28, 1895, abstract at rsnz.natlib.govt.nz/volume/rsnz_28/rsnz_28_00_006110.html. See also: On seasonal time, vol. 31, 1898, rsnz.natlib.govt.nz/volume/rsnz_31/rsnz_31_00_008570.html.

2. "Many studies have shown that working odd hours is associated with health problems …": R. M. Griffin, The health risks of shift work, WebMD, www.webmd.com/sleep-disorders/excessive-sleepiness-10/shift-work.

Chapter Twenty-One: *Science and Destiny*

1. "A few people — solipsists — have argued that maybe it doesn't." Society of Solipsists, www.webcitation.org/6ZZBNVxTd.

2. "Solipsism holds that the universe could be merely in the mind and imagination of the observer …": M. D. Crawford, Dissociation, wayback.archive.org/web/*/geometricvisions.com/schizoaffective-disorder/dissociation.html.

3. "When this network is suppressed, as by psilocybin, the active ingredient of hallucinogenic 'magic mushrooms,' strange things are vividly perceived, but they don't really exist." R. L. Carhart-Harris, et al.,

Neural correlates of the psychedelic state as determined by fMRI studies with psilocybin, *Proceedings of the National Academy of Science (PNAS)*, Jan. 23, 2012, www.pnas.org/content/early/2012/01/17/1119598109.full.pdf+html.

4. *"Science and induction"*: For more information, see P. Godfrey-Smith, *Theory and Reality*, University of Chicago Press, 2003.

5. "This is called 'empiricism.'" M. Shermer, The principle of empiricism, or see for yourself, in J. Brockman, ed., *This Explains Everything*, Edge Foundation Inc., pp. 399–401, 2013.

6. "... the principle: *anything goes*." P. Feyerabend, *Against Method*, 1975, p. 28 of 1978 edition pub. by Verso, ISBN 0860917002.

7. "... *all methodologies ... have their limits*." Preceding reference, p. 32.

8. **"Science vs. pseudoscience"**: See for example: (1) M. Gardner, *Fads & Fallacies in the Name of Science*, Dover, 1957; (2) D. Kossy, *Kooks*, Feral House, 1994.

9. "There exists a continually expanding body of such publications, but there would be even more if this body started including items that can be important in advancing the field but are currently under-disseminated ...": This concept helps motivate some (but not enough) forums like the *Journal of Negative Results in Biomedicine* (jnrbm.com), and *BMC Research Notes* (www.biomedcentral.com/bmcresnotes). See e.g. K. Kelly, Speculations on the future of science, www.kk.org/thetechnium/archives/2006/03/speculations_on_1.

10. "to boldly go where none have gone before": Based on the Star Trek theme song.

Chapter Twenty-Two: *TheTeeming Cities of Mars*

1. "... the first permanent, self-sustaining colony on *Mars*." P. J. Boston, ed., *The Case for Mars; Proceedings of the Conference*, University of Colorado, Boulder, Apr. 29–May 2, 1981. Several subsequent *The Case for Mars* conferences have been held since then. A book by the same title was published in 1997 and revised in 2011.

2. "Mars, the red planet, captor of the imagination ...": R. Crossley, *Imagining Mars: A Literary History*, Wesleyan University Press, 2011.

3. "Mars One starting taking applications in 2013 for a 2023 mission ..." See mars-one.com.

4. "... analysis showed that their life support technology is not yet reliable." J. Chu, Assessment of the technical feasibility of the proposed Mars One mission, *Phys.org*, October 14, 2014, phys.org/news/2014-10-technical-feasibility-mars-mission.html.

5. "Manufacturing equipment would also need to be brought along capable of producing the materials needed for the colony ...": M. Kayser, SolarSinter Project, www.markuskayser.com. This project demonstrates the feasibility of a small, lightweight, solar powered glass manufacturing facility that could be adapted for transport to Mars.

6. "Another piece of bad news you should be informed of is the roughly 2–4% risk of mission failure." J. Foust, Weighing the risks of human spaceflight, *The Space Review*, July 21, 2003, www.thespacereview.com/article/36/1.

7. "... either you get it or you don't ...": H. Macdonald, J. Jesko, H. Karar and L. Effron, Signing up for a mission to Mars, and planning to never return, *ABC News*, March 5, 2015, www.webcitation.org/6WoZ4fz0o .

Chapter Twenty-Three: *Big Ice*

1. "... enabled him to lay the foundations of his life's work: understanding the interaction between Earth's motion in space and the periods in the distant past when glaciers descended from the North Pole — times popularly known as 'ice ages.'" M. Milankovitch, O pitanju astronomskih teorija ledenih doba (Astronomical theory of periods of increased glaciation), University of Zagreb, 1914, legati.matf.bg.ac.rs/milankovic/book.wafl?book=o_pitanju_ledenih_do ba (in Serbian).

2. "The congealed venomous streams continued to send out frost." K. Mortensen, trans. by A. C. Crowell, *A Handbook of Norse Mythology*, ch. 1, p. 49, Thomas Y. Crowell Co., e.g. www.scribd.com/doc/54300132/2/I-HOW-THE-WORLD-WAS-CHEATED.

3. "123,000 years is three 41,000-year obliquity cycles." R. N. Drysdale, et al., Evidence for obliquity forcing of glacial termination II, *Science*, vol. 325, pp. 1527–1531, 2009, sciencemag.org/content/325/5947/1527.full.html.

4. "… averaging about 100,000 years and thus by coincidence giving a false impression of being driven by the 100,000 year eccentricity cycle." P. Huybers and C. Wunsch, Obliquity pacing of the late Pleistocene glacial terminations, *Nature*, 2005, vol. 434, pp. 491–494, www.people.fas.harvard.edu/~phuybers/Doc/pace_nature2005.pdf.

5. "Something — perhaps very gradual cooling over the past few million years (despite the current global warming spike caused by human activity) is causing glacial retreat events to skip obliquity cycles (Figure 10)." P. J. Huybers, Early Pleistocene glacial cycles and the integrated summer insolation forcing, *Science*, 2006, vol. 313, pp. 508–511, nrs.harvard.edu/urn-3:HUL.InstRepos:3382981.

6. "… vast glacial coverings cannot retreat unless they build up first, and the buildup occurs slowly compared to the more dramatic retreat events." E.g. R. Rohde, Ice age temperature changes, Global Warming Art, www.globalwarmingart.com/wiki/File:Ice_Age_Temperature_Rev_png.

7. Figure 10. Based on (1) R. Rohde, Five million years of climate change from sediment cores, Global Warming Art, en.wikipedia.org/wiki/File:Five_Myr_Climate_Change.svg, and (2) L. E. Lisiecki and M. E. Raymo, A Pliocene-Pleistocene stack of 57 globally distributed benthic $\delta^{18}O$ records, *Paleoceanography,* 2005, vol. 20, PA1003, lorraine-lisiecki.com/LisieckiRaymo2005.pdf.

8. Figure 11. Based on en.wikipedia.org/wiki/File:SummerSolstice65N-future.png.

Chapter Twenty-Five: **New Plant Paradigms**

1. "This is called an endospore …": See www.youtube.com/watch?v=NAcowliknPs for how endospores are made.

2. "Without artificially fixed nitrogen in fertilizer, our world would be a very different place." J. W. Erisman, M. A. Sutton, J. Galloway, Z. Klimont,

and W. Winiwarter, How a century of ammonia synthesis changed the world, *Nature Geoscience*, vol. 1, pp. 636–639, 2008, www.nature.com/ngeo/journal/v1/n10/pdf/ngeo325.pdf.

3. "A project announced in 2012 to do just that for food plants like wheat, corn, and rice could help farmers worldwide." British GM crop scientists win $10m grant from Gates, BBC News, July 15, 2012, www.bbc.co.uk/news/science-environment-18845282.

4. "Processed fungus protein called mycoprotein, sold in grocery stores, tastes like chicken already." See www.quorn.com.

5. "… polyploidy induction (increasing the number of copies of each chromosome) can help." E.g. (i) G. D. Ascough, J. van Staden, and J. E. Erwin, Effectiveness of colchicine and oryzalin at inducing polyploidy in *Watsonia lepida* N.E. Brown, *HortScience*, vol. 43, no. 7, pp. 2248–2251, Dec. 2008, hortsci.ashspublications.org/content/43/7/2248; (ii) M. C. De Pablo et al., Effect of induced polyploidy on some characteristics of seed production and quality in *Lotus glaber* Mill., *Lotus Newsletter*, vol. 34, pp. 45–50, 2004, www.inia.org.uy/sitios/lnl/vol34/depablo.pdf; (iii) F. Bretagnolle, J. D. Thompson, and R. Lumaret, The influence of seed size variation on seed germination and seedling vigor in diploid and tetraploid *Dactylis glomerata* L., *Annals of Botany*, vol. 76, no. 6, pp. 607–615, 1995, aob.oxfordjournals.org/content/76/6/607.

6. "Babies are puzzled by fruit …. But within a day or two practically all decide they love it." B. Spock and M. B. Rothenberg, *Dr. Spock's Baby and Child Care, 8th Edition*, 2004, Simon & Schuster, p. 315, ISBN 0-7434-7667-0.

7. "Coconut flavored pineapples (why aren't they called coconapples?), which already exist …": J. Pearlman, Scientists create 'coconut-flavoured' pineapple, *The Telegraph*, Dec. 5, 2012, www.telegraph.co.uk/earth/agriculture/geneticmodification/9723258/Scientists-create-coconut-flavoured-pineapple.html.

8. "Species of zooxanthelae can have 17 known distinct forms." O. A. Jones and R. Endean, editors, *Biology and Geology of Coral Reefs, Volume II: Biology 1,* Academic Press, 1973, p.98, ISBN 0124144756, www.amazon.com/Biology-Geology-Coral-Reefs-II/dp/0124144756.

9. "quite different morphologically": J. P. de Magalhães, et al., Genome-environment interactions that modulate aging: Powerful targets for drug discovery, *Pharmacological Reviews*, Jan. 2012, vol. 64, No. 1, pp. 88–101, pharmrev.aspetjournals.org/content/64/1/88.

10. "… the most Earth-similar planet known," Wikipedia has a convenient list at en.wikipedia.org/wiki/List_of_potentially_habitable_exoplanets.

11. "Genetic engineering can already help plants extract heavy metals like cadmium, lead, and copper from the soil …": P. Kotrba, J. Najmanova, T. Macek, T. Ruml, and M. Mackova, Genetically modified plants in phytoremediation of heavy metal and metalloid soil and sediment pollution, *Biotechnology Advances*, Nov.–Dec. 2009, vol. 27, issue 6, pp. 799–810.

12. "… the potential for toxicity to plants and the high chemical reactivity of aluminum …": E. Delhaize and P. R. Ryan, Aluminum toxicity and tolerance in plants, *Plant Physiology*, 1995, vol. 107, pp. 315-321, www.plantphysiol.org/content/107/2/315.full.pdf.

13. "The organism is named *Pyrococcus furiosus* …": G. Fiala and K. O. Stetter, Pyrococcus furiosus sp. nov. represents a novel genus of marine heterotrophic archaebacteria growing optimally at 100°C, *Archives of Microbiology*, June 1986, vol. 145, no. 1, pp. 56–61.

14. "By splicing such genes into a plant in the cabbage family called *Arabidopsis thaliana* … this … plant can be made much more resistant to heat and dryness." W. F. Boss and A. M. Grunden, Redesigning living organisms to survive on Mars, *NASA Institute for Advance Concepts Annual Meeting*, 2006, www.niac.usra.edu/files/library/meetings/annual/oct06/1194Boss.pdf.

15. "The same genes have also been spliced into tomatoes, which could help feed future colonists." W. Boss, Dec. 8, 2012, www.cals.ncsu.edu/plantbiology/BossLab/hfiles/overview.html.

16. "… the Arctic Apple … does not turn brown when sliced," USDA announces deregulation of non-browning apples, US Department of Agriculture — Animal and Plant Health Inspection Service, www.aphis.usda.gov/stakeholders/downloads/2015/SA_arctic_apples.pdf.

17. "… genetic modifications …": Petitions for determination of nonregulated status, US Department of Agriculture — Animal and

Plant Health Inspection Service,
www.aphis.usda.gov/biotechnology/petitions_table_pending.shtml.

Chapter Twenty-Six: *Asteroid Apocalypse*

1. "By my calculations, 10 megatons of TNT is about 20 cubic acres (5 cubic hectares) of the stuff ...": Define a cubic acre as the volume of a cube 1 acre on a face, and similarly for a cubic hectare.
2. "It was not an H-bomb, however, but evidently an asteroid tens of meters (perhaps as few as 20) in diameter ...": M. B. E. Boslough and D. A. Crawford, Low-altitude airbursts and the impact threat, *International Journal of Impact Engineering*, Dec. 2008, vol. 35, issue 12, pp. 1441–1448.
3. "I say 'crashed into the atmosphere' because asteroids that approach Earth most commonly do so in the neighborhood of 20,000 miles/hr (32,000 km/h)." Derived from data at neo.jpl.nasa.gov/ca.
4. "In fact on June 1 and 2, 1998, two comets did crash into the Sun ...": Twin comets race to death by fire, NASA Goddard Space Flight Center, June 3, 1998, umbra.nascom.nasa.gov/comets/comet_release.html.
5. "The impacts left dark markings easily visible through telescopes and emitted large bursts of microwaves, X-rays, far ultraviolet, infrared, and radio waves."

   Microwaves: H. O. Vats, M. R. Deshpande, O. P. N. Calla, N. M. Vadher, B. M. Darji, V. Sukumaran, Microwave bursts from Jupiter due to K, N, P2 and S fragments, *Earth, Moon, and Planets*, 1996, vol. 73, pp. 125–132, www.springerlink.com/content/m14h632291434u05/fulltext.pdf.

   X-rays: C. J. Hamilton, Hubble observations shed new light on Jupiter collision, *Views of the Solar System*, www.solarviews.com/eng/levyhst.htm.

   Far ultraviolet: G. E. Ballester, et al., Far-UV emissions from the SL9 impacts with Jupiter, *Geophysical Research Letters*, 1995, vol. 22, no. 17, p. 2425–2428.

Infrared: (1) J. Rosenqvist, et al., Four micron infrared observations of the comet Shoemaker-Levy 9 collision with Jupiter at the Zelenchuk Observatory: Spectral evidence for a stratospheric haze and determination of its physical properties, *Geophysical Research Letters*, 1995, vol. 22, no. 12, pp. 1585–1588; (2) R. W. Carlson, et al., Galileo infrared observations of the Shoemaker-Levy 9 G impact fireball: A preliminary report, *Geophysical Research Letters*, 1995, vol. 22, no. 12, pp. 1557–1560.

Radio waves: S. J. Bolton, R. S. Foster, and W. B. Waltman, Observations of Jupiter's synchrotron radiation at 18 cm during the comet Shoemaker-Levy/9 impacts, *Geophysical research letters*, 1995, vol. 22, no. 13, pp. 1801–1804.

6. "You can actually collect micrometeorites yourself from roof runoff and other sources using a magnet, paper, and microscope …": Just hunt around on the web for pages and videos about how to do it!

7. "Were it to crash it would release 510 megatons of energy." 99942 Apophis (2004 MN4) Earth impact risk summary, neo.jpl.nasa.gov/risk/a99942.html.

8. "Fortunately it then dropped rapidly, ruling out initial fears." J. D. Giorgini, L. A. M. Benner, S. J. Ostro, M. C. Nolan, and M. W. Busch, Predicting the Earth encounters of (99942) Apophis, *Icarus*, 2008, vol. 193, pp. 1–19, neo.jpl.nasa.gov/apophis/Apophis_PUBLISHED_PAPER.pdf.

9. "… Apophis presents less than 1 chance in a million of impact in its later approaches of 2036, 2068, 2076, and 2103." 99942 Apophis (2004 MN4) Earth impact risk summary, neo.jpl.nasa.gov/risk/a99942.html.

10. "… it reached the highest Torino Impact Hazard Scale rating ever recorded …": D. Yeomans, S. Chesley, and P. Chodas, Near-Earth asteroid 2004 MN4 reaches highest score to date on hazard scale, NASA Near Earth Object Program, Dec. 23, 2004, neo.jpl.nasa.gov/news/news146.html.

11. "There are roughly 900 large (1 kilometer or more in diameter) near Earth objects known." NEO discovery statistics, NASA Near Earth Object Program, neo.jpl.nasa.gov/stats.

12. "Ninety-two were discovered in the year 2000 but there has been a marked trend of decreasing new discoveries in the years since then." NEO discovery statistics, NASA Near Earth Object Program, neo.jpl.nasa.gov/stats.

13. "The probability was estimated at 1 in 20,000 as of June 26, 2015," 29075 (1950 DA) Earth impact risk summary, NASA, neo.jpl.nasa.gov/risk/a29075.html.

14. "… impacts of at least Tunguska size are believed to occur on the order of once every thousand years …": P. Brown, R. E. Spalding, R. O. ReVelle, E. Tagliaferri, and S. P. Worden, The flux of small near-Earth objects colliding with the Earth, *Nature*, Nov. 21, 2002, vol. 420, pp. 294–296.

15. "… asteroids as big as the one responsible for the demise of the dinosaurs (or larger) are thought to crash into the Earth only once every 200 million years or so." C. R. Chapman, Meteoroids, meteors, and the near-Earth object impact hazard, *Earth Moon Planet*, 2008, vol. 102, pp. 417–424.

16. "… major species extinction events may be occurring every 26-27 million years." M. R. Rampino and K. Caldeira, Periodic impact cratering and extinctions events over the last 260 million years, *Monthly Notices of the Royal Astronomical Society*, vol. 454, p. 3480–3484, 2015. http://mnras.oxfordjournals.org/content/454/4/3480.full.

17. "On the other hand, impacts with energies of about 2 pounds of TNT … occur at the rate of roughly 3 a day." C. R. Chapman, Meteoroids, meteors, and the near-Earth object impact hazard, *Earth Moon Planet*, 2008, vol. 102, pp. 417–424.

18. "The Association of Space Explorers, billed as the 'international professional organization of astronauts and cosmonauts …'": R. L. Schweickart, T. D. Jones, F. von der Dunk, and S. Camacho-Lara, Asteroid threats: A call for global response, p. 34, Association of Space Explorers, 2008, www.space-explorers.org/ATACGR.pdf.

19. "… protracted debate … can lead to inaction; evacuation of the impact site may then be our only option." Association of Space Explorers, exact quote is no longer online, but see similar statements in www.space-explorers.org/ATACGR.pdf.

20. *Spaceguard Survey: Report*: D. Morrison, Spaceguard Survey: Report of the NASA international near-Earth object detection workshop, Jet Propulsion Laboratory, Cal. Inst. of Tech., Pasadena, CA, 1992.

21. "This objective seems likely to be achieved a few years late." Near-Earth object survey and deflection analysis of alternatives (report to congress), National Aeronautics and Space Administration (NASA), 2007, neo.jpl.nasa.gov/neo/report2007.html.

22. "... a rocky asteroid 100 feet (30 m) in diameter could easily be around 600,000 tons ...": C. Q. Choi, Small asteroids pose big new threat, Space.com, December, 2007, www.space.com/4760-small-asteroids-pose-big-threat.html.

23. "A variety of pushing strategies have been devised." Near-Earth object survey and deflection analysis of alternatives (report to congress), National Aeronautics and Space Administration (NASA), 2007, neo.jpl.nasa.gov/neo/report2007.html.

24. "Nuclear explosives are much stronger and according to NASA would be more effective." See preceding ref.

25. "... the Makeyev Rocket Design Bureau in Russia announced plans to repurpose old ICBMs ... into asteroid killers." R. Browne, Russia wants to modify cold war missiles to destroy asteroids, *CNN*, Feb. 19, 2016, www.cnn.com/2016/02/19/politics/russia-icbm-asteroid-killer.

26. "... promoting nuclear explosives could increase risks from them here on Earth," D. Birch, The plans to use nuclear weapons to blow up incoming asteroids, *The Atlantic*, Oct. 16, 2013, www.theatlantic.com/technology/archive/2013/10/the-plans-to-use-nuclear-weapons-to-blow-up-incoming-asteroids/280593.

27. "... you can arrange a Tunguska site tour." See, for example, www.sibtourguide.com/tours/tunguska_explosion_site or www.webcitation.org/6Zcp7gfJZ.

Chapter Twenty-Seven: *Sic Transit Humanitas* — *The Transcent of Man*

1. "*Sic Transit Humanitas*": Inspired by the famous Latin phrase *sic transit gloria mundi*.

2. **"The Transcent of Man"**: By analogy to (1) C. Darwin, *The Descent of Man*; and (2) J. Bronowski, *The Ascent of Man*.

3. "If you happen to have an Andean background ...": J. K. Pritchard, How we are evolving, *Scientific American*, Oct. 2010.

4. "... if you have a Scandinavian or East African dairy farming background ...": see preceding reference.

5. "This was true until the skin lightening A111T mutation of the SLC24A5 gene spread throughout the population." V. A. Canfield, et al., Molecular phylogeography of a human autosomal skin color locus under natural selection, *G3: Genes, Genomes, Genetics*, Nov. 2013, vol. 3, no. 11, pp. 2059–2067, www.g3journal.org/content/3/11/2059.full.

6 "He was revealed to most likely have had relatively dark skin and blue eyes." I. Olalde, et al., Derived immune and ancestral pigmentation alleles in a 7,000-year-old mesolithic European, *Nature*, Jan. 26, 2014, vol 507, pp. 225–228, www.nature.com/nature/journal/v507/n7491/full/nature12960.html, see also www.sciencedaily.com/releases/2014/01/140126134643.htm.

7. "... deliberate selective breeding could potentially produce, in just a few generations, super-athletes, super-geniuses, super-conversationalists, super-sumo wrestlers (already done), and so on." R. Dawkins, *The Greatest Show on Earth: The Evidence for Evolution*, Free Press, 2009.

8. "... it is thought to have been inherited from a child of a union with a Denisovan." E. Huerta-Sánchez, et al., Altitude adaptation in Tibetans caused by introgression of Denisovan-like DNA, *Nature*, 2014, issue 2014/07/02/online, non-paywall summary at www.sciencedaily.com/releases/2014/07/140702131738.htm.

9. "Apparently more closely related to people than any other species at the time, these short, hobbit-like hominids ...": D. Kubo, R. T. Kono, and Y. Kaifu, Brain size of *Homo floresiensis* and its evolutionary implications, Proceedings of the Royal Society B, 2013, vol. 280, no. 1760, rspb.royalsocietypublishing.org/content/280/1760/20130338.

10. "... survived at least until 12,000 years ago ...": M. J. Morwood, et al., Further evidence for small-bodied hominins from the Late Pleistocene of Flores, Indonesia, *Nature*, Oct. 13, 2005, vol. 437, no. 7061, pp.

1012–1017,
www.ncbi.nlm.nih.gov/pubmed?term=16229067.

11. "It turns out that the paracingulate sulcus (PCS), a brain structure, correlates with and likely confers just such an ability." M. Buda, A. Fornito, Z. M. Bergstrom, and J. S. Simons, A specific brain structural basis for individual differences in reality monitoring, *The Journal of Neuroscience*, Oct. 5, 2011, vol. 31, no. 40, pp. 14308–13.

12. "Chimpanzees, interestingly, do not have paracingulate sulci." M. De Haan and M. H. Johnson, *The Cognitive Neuroscience of Development*, Psychology Press, 2003.

13. "Bones showing damage from stone tools suggest that *antecessor* practiced cannibalism and used tools, sometimes at the same time." Y. Fernández-Jalvo, J. Carlos Díez, I. Cáceres, and J. Rosell, Human cannibalism in the Early Pleistocene of Europe (Gran Dolina, Sierra de Atapuerca, Burgos, Spain), *Journal of Human Evolution*, Sept.– Oct. 1999, vol. 37, no. 3–4, pp. 591–622, www.sciencedirect.com/science/article/pii/S004724849990324X.

14. "About 85 million years ago, primates split off ...": H. J. Chatterjee, S. Y. W. Ho, I. Barnes, and C. Groves, Estimating the phylogeny and divergence times of primates using a supermatrix approach, *BMC Evolutionary Biology*, 2009, vol. 9, no. 259, www.biomedcentral.com/1471-2148/9/259.

15. "Apes diverged from monkeys about 32 million years ago." M. E. Steiper, N. M. Young, and T. Y. Sukarna, Genomic data support the hominoid slowdown and an Early Oligocene estimate for the hominoid-cercopithecoid divergence, *Proceedings of the National Academy of Science (PNAS)*, Dec. 7, 2004, vol. 101, no. 49, pp. 17021– 6.

16. "The great apes, a subgroup of apes technically called hominids, came along around 19 million years ago." M. E. Steiper and N. M. Young, Primates, in S. B. Hedges and S. Kumar, *The Timetree of Life*, Oxford University Press, 2009, pp. 482–486.

17. "This ancient species is thus called Nakalipithecus." Y. Kunimatsu, et al., A new Late Miocene great ape from Kenya and its implications for the origins of African great apes and humans, *Proceedings of the*

National Academy of Sciences (*PNAS*), Dec. 2007, vol. 104, no. 49, pp. 19220–5, www.ncbi.nlm.nih.gov/pmc/articles/PMC2148271.

18. "Gorillas split off from our common ancestors about 7 million years ago." S. L. Robson and B. Wood, Hominin life history: Reconstruction and evolution, *Journal of Anatomy*, Apr. 2008, vol. 212, no. 4, pp. 394–425, www.ncbi.nlm.nih.gov/pmc/articles/PMC2409099.

19. "The progenitor of chimps and bonobos split off 5–7 million years ago." S. Kumar, A. Filipski, V. Swarna, A. Walker, and S. B. Hedges, Placing confidence limits on the molecular age of the human-chimpanzee divergence, *Proceedings of the National Academy of Sciences* (*PNAS*), Dec. 27, 2005 vol. 102 no. 52, pp. 18842–18847, www.pnas.org/content/102/52/18842.full.

20. "That progenitor species enjoyed considerable success over time, spawning almost 2 dozen separate identifiable human-like species ...": Table 5 of S. L. Robson and B. Wood, Hominin life history: Reconstruction and evolution, *Journal of Anatomy*, Apr. 2008, vol. 212, no. 4, pp. 394–425, www.ncbi.nlm.nih.gov/pmc/articles/PMC2409099.

21. "One survey puts elephants and whales at 1.3× and 1.8× respectively." G. Roth and U. Dicke, Evolution of the brain and intelligence, *Trends in Cognitive Sciences*, May 2005, vol. 9, no. 5, pp. 250–257.

22. "A dramatic process of brain enlargement in our past began approximately two million years ago." M. A. Hofman, Human brain evolution: design without a designer, *Heredity*, 2007, vol. 20, pp. 62–67.

23. "The function of the FOXP2 gene has been investigated using mice by substituting the human version for the original mouse version and observing how this affects the mice." S. Reimers-Kipping, W. Hevers, S. Pääbo, and W. Enard, Humanized Foxp2 specifically affects cortico-basal ganglia circuits, *Neuroscience*, Feb. 2011, vol. 175, pp. 75–84, dx.doi.org/10.1016/j.neuroscience.2010.11.042.

Chapter Twenty-Eight: *Floating Prairies of the Seas*

1. "… when cooked, the young leaves, stems, and flotation bladders are edible." See e.g. www.youtube.com/watch?v=V1kkn5Sz4MI.
2. "Check local regulations for legal restrictions before transporting to your dinner table." houstonwildedibles.blogspot.com/2008/10/water-hyacinth.html.
3. "Figure 12. Water hyacinths …": Agricultural Research Service, US Dept. of Agriculture, see e.g. en.wikipedia.org/wiki/File:Water_hyacinth.jpg.
4. "by the wind sailor." *Velella velella* - By the wind sailor, SIMoN Sanctuary Integrated Monitoring Network, www.webcitation.org/6Zcq0LGb7.
5. "Indeed, in a 2012 report, iron compounds were dispersed in ocean water, stimulating growth of masses of diatom-rich plankton." V. Smetacek, et al., Deep carbon export from a southern ocean iron-fertilized diatom bloom, *Nature*, July 19, 2012, vol. 487, issue 7407, pp. 313–319.
6. "To get an estimate, suppose a root filament has a dry weight roughly the same as a human hair, about 0.05 milligrams per cm." M. Legrand, C. Passos, D. Mergler, and H. Chan, Biomonitoring of mercury exposure with single human hair strand, *Environmental Science and Technology*, 2005, vol. 39, pp. 4594–4598, www.unites.uqam.ca/gmf/caruso/doc/caruso/passos/legrand_2005.pdf.

Chapter Twenty-Nine: *Get Ready for the Greenish Revolution*

1. "Thus unless human population increase ends across the board as a result of contraceptives, urbanization, and other voluntary checks …": S. Brand, *Whole Earth Discipline*, Penguin Books, 2010.
2. "In the UK, agricultural employment declined …": 20 year plus trendwatch, UK Agriculture, www.ukagriculture.com/farming_today/20year_plus_trends.cfm.

3. "... in the US the percentage of the labor force devoted to agriculture ...":
E. Nosal and M. Shenk, Is manufacturing going the way of agriculture?
Federal Reserve Bank of Cleveland,
www.clevelandfed.org/research/trends/2007/0307/02ecoact.cfm.

4. "... in Australia the percentage decreased ...": L. Lu and D. Hedley, The
impact of the 2002–03 drought on the economy and agricultural
employment, Australian Government Treasury,
www.webcitation.org/6ZZgn8Dkv.

5. "... obesity is a problem in all three countries." F. Sassi, M. Devaux, M.
Cecchini, and E. Rusticelli, The obesity epidemic: Analysis of past and
projected future trends in selected OECD countries, *OECD Health
Working Papers*, 2009, no. 45, OECD publishing,
www.webcitation.org/6ZZgxZ7Q8.

6. "... in terms of the quantity of plant and animal tissue produced ...": E. O.
Wilson, *The Future of Life*, Random House, 2002.

7. "... the greater the diversity of species ...": E. O Wilson (2002).

8. "analog food mill" ... "square strip of heavy paste" ... "chuffing": W.
Macfarlane, Free vacation, *Analog*, Oct. 1967, vol. LXXX, no. 2, pp.
114–125.

9. "... a process for converting indigestible cellulose from arbitrary plants
into healthful and nutritious amylose starch." C. You, et al., Enzymatic
transformation of nonfood biomass to starch, *Proceedings of the
National Academy of Sciences of the United States of America*, Mar.
2013, vol. 110, no. 18, pp. 7182–7187,
www.pnas.org/content/110/18/7182.full.

10. "... edible landscaping." Lots of information exists on the web about
how to do this. The modern movement began with R. Creasy, *The
Complete Book of Edible Landscaping*, Sierra Club Books, 1982.

11. "... Intestinal Fortitude project." M. Donlon, Intestinal Fortitude,
DARPA,
web.archive.org/web/*/www.darpa.mil/dso/thrust/biosci/intefort.htm,
www.wired.com/dangerroom/2007/03/darpa_the_penta.

12. "... coral, sea slugs, and even giant clams can." D. Yellowlees, T. A.
Rees, and W. Leggat, Metabolic interactions between algal symbionts

and invertebrate hosts, *Plant Cell & Environment*, May 2008, vol. 31, no. 5, pp. 679–694.

13. "So do some large snails." T. Berner, A. Wishkovsky, and Z. Dubinsky, Endozoic algae in shelled gastropods — a new symbiotic association in coral reefs? I. Photosynthetically active zooxanthellae in Strombus tricornis, *Coral Reefs*, 1986, vol. 5, pp. 103–106.

14. "Upon loss of the theca, the alga assumes an irregularly shaped form. Fingerlike processes of the algal cells penetrate between adjacent animal cells." J. L. Oschman, *Journal of Phycology*, Sept. 1966, vol. 2, issue 3, pp. 105–111. Also R. E. Lee, *Phycology* (3rd ed.), Cambridge University Press, 1999: "Upon loss of the theca, the alga assumes an irregularly shaped form, with fingerlike processes of the algal cells penetrating between adjacent animal cells."

15. "The chloroplasts are thought to have once been independent organisms …": J. W. Kimball, Endosymbiosis and the origin of eukaryotes, *Kimball's Biology Pages*, 2009, users.rcn.com/jkimball.ma.ultranet/BiologyPages/E/Endosymbiosis.ht ml.

16. "The Sacoglossa, a group of about 300 species of sea slug, are sometimes called "solar-powered sea slugs" because they steal chloroplasts …": M. E. Rumpho, E. J. Summer, and J. R. Manhart, Solar-powered sea slugs. Mollusc/algal chloroplast symbiosis, *Plant Physiology*, May 2000, vol. 123, no. 1, pp. 29–38, www.plantphysiol.org/content/123/1/29.

*Chapter Thirty: **Accelerating Evolution***

1. "… a unitary bird, with its strong, lightweight, complex feathered surface, on the order of 700× denser than the air …": D. M. Hamershock, T. W. Seamans, and G. E. Burnhardt, Determination of body density for twelve bird species. Technical report WL-TR-93-3049, Wright Laboratory, Wright-Patterson Air Force Base, Ohio, 1993, www.dtic.mil/cgi-bin/GetTRDoc?AD=ADA266452. The midpoint of the intersection of the ranges of their three bird density measurement methods is 0.899 g/cm^3.

2. "… microscopic nanobacteria about 200 nanometers across and first photographed in detail in 2015," First detailed microscopy evidence of nanobacteria at the lower size limit of life, *Kurzweil Accelerating Intelligence News*, March 9, 2015, www.kurzweilai.net/irst-detailed-microscopy-evidence-of-nanobacteria-at-the-lower-size-limit-of-life.

3. "The world's number of different kinds of organisms has been increasing, on average, for hundreds of millions of years …": (1) A. V. Markov and A. V. Korotayev, Phanerozoic marine biodiversity follows a hyperbolic trend, *Paleoworld*, Dec. 2007, vol. 16, issue 4, pp. 311–318; (2) www.mun.ca/biology/scarr/Extinction_rates.html; (3) en.wikipedia.org/wiki/File:Phanerozoic_Biodiversity.svg.

4. "These plunges are caused by dramatic mass extinctions, which have a curious tendency to occur roughly 62 million years apart." R. A. Rohde and R. A. Muller, Cycles in fossil diversity, *Nature*, Mar. 10, 2005, vol. 434, pp. 208–210.

5. "Genetic markers consistent with this in fact have a periodicity component of 61 million years." G. Ding, J. Kang, Q. Liu, T. Shi, G.. Pei, and Y. Li, Insights into the coupling of duplications events and macroevolution from an age profile of animal transmembrane gene families, *PLoS Computational Biology*, 2006, vol. 2, no. 8, e102.

6. Figure 13. (1) J. J. Sepkoski, Jr., A kinetic model of Phanerozoic taxonomic diversity, III. Post Paleozoic families and mass extinctions, *Paleobiology*, 1984, vol. 10, pp. 246–267; (2) A. I. Miller and M. Foote, Calibrating the Ordovician radiation of marine life — implications for Phanerozoic diversity trends, *Paleobiology*, 1996, vol. 22, pp. 304–309.

7. "Heteropatry ('different fatherland')": W. M. Getz and V. Kaitala, Ecogenetic models, competition, and heteropatry, *Theoretical Population Biology*, 1989, vol. 36, pp. 34–58.

8. "It occurs most commonly through heteropatry and may be mediated genetically by processes like 'reinforcement' and 'chromosomal inversion.'"

    Reinforcement: D. Ortiz-Barrientos, B. A. Counterman, and M. A. F. Noor, The genetics of speciation by reinforcement, *PLoS Biology*, 2004, vol. 2, no. 12:e416, www.plosbiology.org.

Chromosomal inversion: M. A. F. Noor, K. L. Grams, L. A. Bertucci, and J. Reiland, Chromosomal inversions and the reproductive isolation of species, *Proceedings of the National Academy of Science* (*PNAS*), 2001, vol. 98, no. 21, pp. 12084–12088, www.pnas.org/content/98/21/12084.full.

9. "Darwin, for example, observed that 'offspring of each species will try … to seize … diverse places in the economy of Nature …. Each new variety or species, when formed, will generally take the place of … its less well-fitted parent.'" W. M. Getz and V. Kaitala, Ecogenetic models, competition, and heteropatry, *Theoretical Population Biology*, 1989, vol. 36, pp. 34–58.

10. "The same spot will support more life if occupied by very diverse forms." P. H. Barrett, P. J. Gautey, S. Herbert, D. Kohn, and S. Smith, eds., *Charles Darwin's Notebooks, 1836–1844*, Cambridge University Press, 1987. Numerous other citations may be found by querying a web search engine with this quote from Darwin.

11. "The language also has multiple levels of meaning …": D. Berleant, M. White, E. Pierce, E. Tudoreanu, A. Boeszoermenyi, Y. Shtridelman, and J. C. Macosko, The genetic code — more than just a table, *Cell Biochemistry and Biophysics*, 2009, vol. 55, no. 2, pp. 107–116.

12. "Biomass living in clouds today has the 'potential to increase by as much as 20% per day' …": B. Christner, Cloudy with a chance of microbes, *Microbe Magazine*, Feb. 2012, vol. 7, no. 2, pp. 70–75, www.webcitation.org/6ZZhfSWQC.

13. "Biomass living in clouds today … is estimated to metabolize in the neighborhood of '1 million tons of organic carbon per year.'" P. Amato, Clouds provide atmospheric oases for microbes, *Microbe Magazine*, Mar. 2012, vol. 7, no. 3, pp. 119–123, www.webcitation.org/6ZZhrjdSG.

14. "The bacteria Erwinia caratovora is just one species that spends part of its life cycle in the clouds." E. Newman, University of Wyoming scientist looks for bacteria in clouds, Casper Star Tribune, May 4, 2012, trib.com/news/state-and-regional/university-of-wyoming-scientist-looks-for-bacteria-in-clouds/article_c041805a-8a33-57c8-8da4-e3103cab0be7.html.

15. "Some strains of Pseudomonas syringae, for example, are coated with proteins containing a 'magic' 8 amino acid sequence …": Cloudy with a chance of microbes, *Microbe Magazine*, Feb. 2012, vol. 7, no. 2, pp. 70–75, www.webcitation.org/6ZZhfSWQC.

16. "Such bacteria are called ice nucleators (INs) and are thought to seed clouds to encourage precipitation." C. Dell'Amore, Rainmaking bacteria ride clouds to "colonize" Earth? *National Geographic News*, Jan. 12, 2009, news.nationalgeographic.com/news/2009/01/090112-clouds-bacteria.html.

17. "The significance of this to weather and climate is unknown but perhaps high: Clouds typically hold ten thousand microbes per cubic meter." C. Dell'Amore (2009).

18. "… clouds typically hold ten thousand microbes per cubic meter. In fact a surprising one third of ice crystals in clouds were found to be nucleated biologically …": Cloudy with a chance of microbes, *Microbe Magazine*, Feb. 2012, vol. 7, no. 2, pp. 70–75, www.webcitation.org/6ZZhfSWQC.

19. "This stress will likely start reducing diversity of life …": K. Caldeira and J. F. Kasting, The life span of the biosphere revisited, *Nature*, 24/31 Dec., 1992, vol. 360, pp. 721–723.

20. "Too much heat is also a biosphere stressor, and that will also tend to reduce life," S. Franck, et al., Causes and timing of future biosphere extinctions, *Biogeosciences*, 2006, vol. 3, pp. 85–92.

21. "This mechanism is so robust it permits transfer of genes between individuals of different species." C. Dahlberg, et al., Interspecies bacterial conjugation by plasmids from marine environments visualized by gfp expression, *Molecular Biology and Evolution*, 1998, vol. 15, no. 4, pp. 385–390.

22. "One such gene appears to be the SRGAP2 gene, which is partially duplicated in humans, the partial copy being called SRGAP2C." C. Charrier, et al., Inhibition of SRGAP2 function by its human-specific paralogs induces neoteny during spine maturation, *Cell*, 2012, vol. 149, issue 4, pp. 923–935. Reviewed in: How a gene copy helped our brains become 'human,' *Kurzweil Accelerating Intelligence News*,

May 8, 2012, www.kurzweilai.net/how-a-gene-copy-helped-our-brains-become-human.

23. "Russian biologist Ilya Ivanov (1870–1932) led the effort in the 1920s, though not at the behest of Soviet dictator Josef Stalin as has been colorfully claimed." B. Dunning, Stalin's human-ape hybrids, Skeptoid Media, Inc., Aug. 17, 2010, skeptoid.com/episodes/4219.

Chapter Thirty-One: *If the Universe As We Know it Ends, When Will it Happen?*

1. "… one could always draw stoic comfort from the possibility that perhaps in the course of time …": S. Coleman and F. de Luccia, Gravitational effects on and of vacuum decay, Physical Review D, June 15, 1980, vol. 21, no. 12, pp. 3305–3315, blogs.discovermagazine.com/cosmicvariance/files/2011/10/cdl.pdf; preprint at www.slac.stanford.edu/pubs/slacpubs/2000/slac-pub-2463.html.

2. "… faster than light travel seems possible if speed is described using equations that permit going 'around' the barrier presented by the exact speed of light." C. Asaro, (1) A luminous future, ch. 5 in D. Broderick, ed., *Year Million*, Atlas & Co., 2008; (2) Complex speeds and special relativity, *American Journal of Physics*, vol. 64, Apr., pp. 421–429.

3. "As pointed out by philosophers of science like Karl Popper, scientific theories cannot be proven absolutely." P. Godfrey-Smith, *Theory and Reality*, University of Chicago Press, 2003.

4. "… we don't know if the state of the universe is a false vacuum and, if it is, whether it will ever transition out of it and thus destroy everything we are familiar with." I give a few citations at www.webcitation.org/6SmoptQja.

5. "Then, a randomly chosen point within the lifespan of our area has a 50% chance of being in the first half of that lifespan and a 50% chance of being in the second half (if the probability of a metastability event is constant over time …)." Actually, we cannot be sure whether the probability of a metastability event is the same each year. That

depends on the unknown morphology of the field energy landscape that the universe's state is traversing as the universe expands, thus potentially modifying its field energy configuration. If the probability increases over time, then a random point in time is likely to be nearer the end of the universe than the beginning. On the other hand, if the probability is decreasing then a random point in time is more likely to be nearer to the birth of the universe than its death.

6. "Your chances of being hit by lightning … are … roughly 1 in 10,000 …": Lightning safety, National Weather Service, web.archive.org/web/20130426052608/http://www.lightningsafety.noaa.gov/medical.htm.

7. "Take precautions during storms!" See, e.g., Personal lightning safety tips, National Lightning Safety Institute, www.lightningsafety.com/nlsi_pls/lst.html.

8. "(probably is?)": M. Yglesias, American democracy is doomed, *Vox*, Oct. 8, 2015, www.vox.com/2015/3/2/8120063/american-democracy-doomed.

9. "As another complicating factor, there is actually a variety of apocalyptic events that can occur, and they should all be appropriately analyzed to determine how concerned we should be." W. Wells, *Apocalypse When? Calculating How Long the Human Race Will Survive*, Springer, 2009.

Chapter Thirty-Two: **Questions**

1. "… the prediction favored by most cosmologists is not Big Rip but rather continued expansion and cooling, the universe ultimately sliding into a "Big Freeze." History of the universe, *PBS*, 2000, www.pbs.org/wgbh/nova/universe/history.html.

2. "That our own "real" universe is more likely a simulation than not has been seriously argued in the philosophy field, and experiments capable of providing evidence for or against this are described by physicists."

   Philosophy: N. Bostrom, Are you living in a computer simulation? See www.simulation-argument.com.

Experiments: S. R. Beane, Z. Davoudi, M. J. Savage, Constraints on the universe as a numerical simulation, *arXiv*, arXiv:1210.1847v2, arxiv.org/abs/1210.1847v2.

3. "… it is known that most species last no longer than a few million years." S. J. Gould, *The Structure of Evolutionary Theory*, Harvard University Press, 2002.

Acknowledgments

This book has benefited greatly from input by numerous parties; I am grateful to each. They include David Alexander, Arnold and Riva Berleant, Keith Bush, CASFWG, Keith Curtis, Joseph Friedlander, Bill Friedman, Maciamo Hay, Tom Howard, David Hutchinson, Jack Jackson, Maryann Karinch, Alan Kauffman, Daniel Levine, Donald Maclean, Shirin Mirlohi, Margaret Morris, the Natural Philosophy Study Group, Steve Nerlich, Megan Neumann, Terri Proksch, Jonathan Simonson, Logan Streondj, Viktor Toth, Tihamer Toth-Fejel, UALR, Washington Academy of Sciences, Willard Wells, Eleanor and George (1928–2012) Wolff, Eve Syrkin Wurtele, Paul Yoder, and others. You probably know why you're here and have my heartfelt thanks! If your name was inadvertently omitted, please let me know. A special mention goes to Eric Klien of the Lifeboat Foundation for his confidence in this book and, in addition, his invaluable editorial and other support.

I also extend my thanks to you, the reader, for your interest in this fascinating subject we call the future. If you would like to submit a suggestion or correction, however large or small, for the next edition, please go to lifeboat.com/ex/book or to the Facebook discussion group, facebook.com/groups/thehumanracetothefuture. It will be received with appreciation.

Index

grass, 45, 46, 217

gravity, 114, 115, 116, 117, 176, 244

Great Britain, 72

Great Dying, 150, 152, 153

green fluorescent protein, 308, 309

greenhouse effect, 65, 108, 149, 234, 341

Greenland, 152

Groucho Marx, 267

gyres, 280

hair, 19, 20, 282, 318, 331, 370

Haldane, J. B. S., 129, 352

hallucinations, 51, 83, 260

hand, 241, 242

happiness, 28, 334

health, 8, 11, 31, 44, 45, 46, 47, 53, 55, 64, 65, 113, 114, 168, 169, 180, 219, 222, 268, 287, 288, 291, 338, 339, 345

health care, 31, 55

heat, 73, 104, 105, 113, 199, 218, 235, 242, 244, 250, 362

Heisenberg uncertainty principle, xii, 135, 136, 137, 138, 139

helmet, 19, 331

heroism, 118

heteropatry, 298, 373

Hillel, 145, 354

HIV, 256

hobbits, 261, 367

holidays, 114

hominids, 259, 265, 266, 367, 368

Homo antecessor, 261

Homo floresiensis, 258, 261

Homo heidelbergensis, 261, 263

Homo sapiens, 140, 143, 265, 308

homosexuality, 131

hormones, 45

horses, 288

Houston, 253

Hudson, G. V., 168, 357

human condition, 88, 93, 272, 273

humankind, 11, 23, 25, 65, 71, 72, 75, 79, 87, 88, 89, 92, 93, 99, 143, 145, 146, 147, 150, 174, 187, 191, 192, 259, 267, 270, 273, 284, 289, 300, 321

humanzees, 310

hummingbirds, 242

Hungary, 195

hunter-gatherer, 64, 65

Hurricane Rita, 253

Hydergine, 9, 326

hydrogen sulfide, xii, 149

hydromel, ix, 8, 9, 15, 326

hypertext, 23, 24, 332, 333

hypnosis, 82

hypotheses, 97, 178, 179

IBM, 87

ice, 84, 105, 111, 119, 151, 195, 201, 202, 203, 205, 206, 207, 223, 235, 245, 303, 304, 345, 359, 374, 375

ice age, 201, 202, 207

ice nucleators, 304

If I am not for myself, who will be for me?, 145, 354

immortal, 73, 342

incentives, x, 19, 28, 54, 59, 62, 73, 99, 229, 288

India, 29, 94

Indian Ocean, 253, 281

indoctrination, 59

Indonesia, 258, 367

induction, 175, 176, 357

About the Cover Artists and Designers

The cover of the book you are reading features one of a collection of fine works by artists associated with Lifeboat. This book is distributed featuring each artist at different times. To view and enjoy all of them online, simply visit the publisher at lifeboat.com/ex/book. Please read more about the artists below.

Catherine Asaro received her PhD in chemical physics from Harvard University and wrote her doctorate in theoretical atomic and molecular physics. For the past twenty years, she has been a novelist with over twenty-five science fiction novels and near future thrillers published, as well as numerous works of shorter fiction and non-fiction. She was elected to two terms as the president of the board of directors for the Science Fiction and Fantasy Writers of America, sand is the president and founder of Starflight Music, an independent record label. She also teaches as a visiting professor in the Department of Physics at the University of Maryland, Baltimore County, and coaches profoundly gifted math students in high school and middle school for top-ranked national and international contests, for example the USA Mathematical Olympiad and the American Regional Mathematics League. She is a member of SIGMA, a think tank of speculative writers and scientists who advise the government regarding future trends affecting national security.

Catherine is a two-time winner of the Nebula Award, sometimes known as the "Oscar of Science Fiction," for her novel *The Quantum Rose* and her novella *The Spacetime Pool*, and she has won numerous other awards for her science fiction, fantasy, and high-tech thrillers. She appears as the Author Guest of Honor, as a futurist and keynote speaker, and as a singer at numerous conferences in the United States and abroad, including for example Guest of Honor at the New Zealand National Science Fiction Convention.

Find out more at lifeboat.com/ex/bios.catherine.asaro.

Design:
Catherine
Asaro

Clock art:
Francesco
De Comité

J. Daniel Batt is a writer, teacher, designer, and developer of community groups. He currently serves as the Art Director at a nonprofit organization in Sacramento, California. He has written and published several short stories including the children's tale "Keaghan in Dreamside" and is the designer of many more. In addition, he is preparing the release of his first full-length novel, *The Young Gods*. Jason is a member of the Lifeboat Foundation, serves on the Advisory Board for the University of Arizona's Consortium for Applied Space Ethics, and was a speaker at the 100 Year Starship Symposiums in 2011 and 2012. He and his family live in northern California.

Find out more at lifeboat.com/ex/bios.jason.batt.

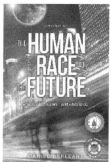

Frank D. Smith is a Lifeboat Foundation staff member and project coordinator of the InfoPreserver skills repository wiki at lifeboat.com/infopreserver (email: infopreserver@lifeboat.com). An "artnician" from Los Angeles who has interwoven art and technology throughout his career, he now makes his home in Bellingham, Washington, USA with his beautiful wife Kimberly and amazing stepdaughter Siri. He hopes to be creating holographic covers for Lifeboat Foundation publications in a hundred years.

Find out more at lifeboat.com/ex/bios.frank.d.smith.

About the Author

Daniel Berleant earned his PhD and MS from the University of Texas at Austin in 1991. His BS is from MIT. He has worked in industry as a software engineer and taught courses at three universities to students ranging from the freshman to advanced graduate levels. Berleant and his students' research has been funded through the National Science Foundation, National Institutes of Health, Corporation for National and Community Service, and other agencies; and through industry including Procter and Gamble, Electricité de France, Invitrogen, and others. Research in his lab has included artificial intelligence, inference under severe uncertainty, web science, energy, text mining, bioinformatics, and technology foresight. Find out more at lifeboat.com/ex/bios.daniel.berleant.

Made in the USA
Lexington, KY
10 July 2016